CONCISE DICTIONARY OF
AMERICAN LITERATURE

MIDCENTURY
REFERENCE LIBRARY

DAGOBERT D. RUNES, Ph.D., *General Editor*

AVAILABLE

Dictionary of American Grammar
and Usage
Dictionary of American Proverbs
Dictionary of Ancient History
Dictionary of the Arts
Dictionary of European History
Dictionary of Etiquette
Dictionary of Foreign Words
and Phrases
Dictionary of Forgotten Words
Dictionary of Last Words
Dictionary of Linguistics
Dictionary of Mysticism
Dictionary of Mythology
Dictionary of Pastoral Psychology
Dictionary of Philosophy
Dictionary of Psychoanalysis
Dictionary of Science and Technology
Dictionary of Sociology
Dictionary of Word Origins
Dictionary of World Literature

Encyclopedia of Aberrations
Encyclopedia of the Arts
Encyclopedia of Atomic Energy
Encyclopedia of Criminology
Encyclopedia of Literature
Encyclopedia of Psychology
Encyclopedia of Religion
Encyclopedia of Substitutes and
Synthetics
Encyclopedia of Vocational Guidance
Illustrated Technical Dictionary
Labor Dictionary
Liberal Arts Dictionary
Military and Naval Dictionary
New Dictionary of American History
New Dictionary of Psychology
Protestant Dictionary
Slavonic Encyclopedia
Theatre Dictionary
Tobacco Dictionary

FORTHCOMING

Beethoven Encyclopedia
Dictionary of American Folklore
Dictionary of the American Language
Dictionary of American Literature
Dictionary of American Maxims
Dictionary of American Names
Dictionary of American Superstitions
Dictionary of American Synonyms
Dictionary of Anthropology
Dictionary of Arts and Crafts
Dictionary of Asiatic History
Dictionary of Astronomy
Dictionary of Child Guidance
Dictionary of Christian Antiquity
Dictionary of Dietetics
Dictionary of Discoveries and Inventions
Dictionary of French Literature
Dictionary of Geography
Dictionary of Geriatrics
Dictionary of German Literature

Dictionary of Hebrew Literature
Dictionary of Judaism
Dictionary of Latin Literature
Dictionary of Law
Dictionary of Mathematics
Dictionary of Mechanics
Dictionary of Mental Hygiene
Dictionary of New Words
Dictionary of Physical Education
Dictionary of Poetics
Dictionary of the Renaissance
Dictionary of Russian Literature
Dictionary of Science
Dictionary of Social Science
Dictionary of Spanish Literature
Encyclopedia of American Philosophy
Encyclopedia of Morals
Personnel Dictionary
Teachers' Dictionary
Writers' Dictionary

LOUISA M. ALCOTT

Concise
Dictionary of
AMERICAN
LITERATURE

Edited by ROBERT FULTON RICHARDS

Philosophical Library :-: New York

PREFACE

This dictionary was prepared with continuing reference to the question, "What is important and meaningful in the study of American literature?" It was designed primarily for the student, although we hope that instructors also will find it useful. It was conceived from the premise that the student's particular needs may be served by a selection and arrangement of material that is quite different from that of the normal reference book.

The dictionary which is planned merely as a source of facts is apt to present them in a consistent pattern, with no thought of the impression they give. Thus we are told that Ezra Pound was born in Idaho (1885) and graduated from Hamilton College (1905). The student, ever eager to interpret facts, assumes that Pound grew up in Idaho, and looks for western influences in his poetry. But Pound was removed to the East at the age of two. This very simple example may illustrate how, in a variety of more complex considerations, the material for the student's reference book must be selected with deliberation. A catalogue of facts also has the disadvantage of making literature appear a storehouse of dry, dead, information, and the student is better served if the facts are presented with some consideration of style.

I presume, and thank God, that American literature is no longer taught as the study of Puritan theology, Colonial politics, and nineteenth-century pragmatism, with a few creative artists included for balance. This is not a dictionary of the main currents of American thought, although the importance of ideology in our literary development is recognized.

I have noted a tendency, among colleagues who have examined this work, to assume that the space devoted to an author is an inflexible measure of his importance or even of his quality. My object was to present material according to its value in the study of American literature, and the quantitative disparity between great and minor authors is accordingly large; however, if an author's work needed little explanation, if scholars had found few problems in his study, or if his life was uneventful, he could be covered in a shorter space than others of lesser rank. Each entry is, in Lincoln's terminology, long enough to reach the ground.

Plot outlines of novels are not included, for I presume that the student is still expected to read the books he studies. Brief statements about the nature and quality of an author's books are presented within each entry, for this is information that the student may need beyond the scope of his reading assignments.

I wish to express my appreciation to Professor Lewis Leary of Columbia University for his suggestions concerning procedure, to Professor Richard Walser of North Carolina State College for preparation of the North Carolina entries, and to Chris Richards for her invaluable editorial assistance.

R. F. R.

Notes concerning the text

Attention is directed to the five surveys, included alphabetically in the text, which may have a special value for the student:

Criticism
Drama
Novel
Poetry
Short Story

Authors are listed alphabetically by surname. Those names or parts of names which were not used by the author are placed in brackets, as: FITZGERALD, F[rancis] Scott [Key]. Authors who are known primarily by their pseudonyms are so listed, with a cross reference to their proper names, so that the reader will not need first to look up "Arp, Bill," and then turn to Charles H. Smith.

Publication dates refer to American editions. Dates given for plays refer to the year of initial production. Pulitzer Prize awards are not dated, since they uniformly are presented in the year following publication or production, and are awarded for the year in which the prize work appeared.

Since the geographical definition, United States, becomes rather cumbersome in repeated usage and serves poorly in the adjectival form, "America" is consistently used to define the United States of America, and apologies are extended to the other North and South American nations.

The following abbreviations are used:

c.	(circa)	about.
fl.	(floruit)	flourished.
i.e.	(id est)	that is to say.
q.v.	(quod vide)	which see.
qq.v.	(quae vide)	both or all of which see.

The use of the q.v. is limited to cross references which will add significant material to the item under discussion.

A

ABBEY, Henry (1842-1911), a New York businessman, in his leisure wrote minor, pleasant poems that were popular in his day. Of the twelve collections of his work published, the best are *Ballads of Good Deeds* (1872) and *Poems* (1879).

ABBOTT, Jacob (1803-79), author of the "Rollo Books" (28 vols., 1834 ff.), was the first writer of juvenile fiction to mix some humanity with his didacticism. A progressive educator, he taught school in many parts of New England and New York, and had both Longfellow and Helen Hunt Jackson for pupils. He published more than 200 works. Lyman Abbott, the rational theologian, was his son.

ADAIR, James (c. 1709-c. 1783), Irish trader who came to South Carolina in 1735, in 1775 wrote *The History of the American Indians*. This account of his adventures among the Chickasaws reasons that the Indians are descended from the ancient Jews, yet contains valuable historical reporting.

ADAMIC, Louis (1899-1951), ran away from his Austrian home at fourteen to avoid a Jesuit training, and came to the United States, where he worked at various jobs until Mencken accepted an article for *The American Mercury* in 1928. His work soon began to appear in other magazines, and in 1931 his labor study, *Dynamite*, was published. *The Native's Return* (to his homeland) was a Book-of-the-Month Club selection in 1934. His work is characterized by sensitivity to the economic and social backgrounds of both his native and his adopted countries.

ADAMS, Andy (1859-1935), a Texas cattle rancher and Colorado miner, achieved a surprising literary quality in his narratives of the cattle industry: *The Log of a Cowboy* (1903), *The Outlet* (1905), *Cattle Brands* (1906), and *Reed Anthony, Cowman: an Autobiography* (1907). In addition to authentic portraits in this highly romanticized field, he gave the social and political backgrounds of the cattle trade.

ADAMS, Franklin P[ierce] (1881-), New York journalist, long a member of the "Information Please" radio panel, imitated the language and style of *Pepys's Diary* in his column, "The Conning Tower," from which was selected the *Diary of Our Own Samuel Pepys* (1935). *The Melancholy Lute* (1936) is a selection of his satiric verse which challenges popular platitudes.

ADAMS, Henry [Brooks] (1838-1918), great grandson of President John Adams, grandson of President John Quincy Adams, and son of the diplomat, Charles Francis Adams, was educated at private Boston schools, at Harvard, and in Germany. During the Civil War he served as private secretary to his father, then Minister to England, which experience Henry described in a letter to his fighting brother, Charles, as "dragging our weary carcasses to balls and entertainments of every description." Other letters show, however, that he thoroughly enjoyed the company of the charming ladies. As a journalist in Washington, 1868-70, he was reported to be "the capital's best dancer." Yielding to family pressure and arrangements, he accepted the position of Assistant Professor of Medieval History at Harvard, 1870, for which he had no apparent qualifications, and he later complained that his students were "disgustingly clever at upsetting me with questions." Yet many of his students became prominent medievalists, and his *Mont Saint-Michel and Chartres* (1904) has become a definitive study of the medieval spirit, or "force," represented by the Virgin. In 1872 he married the heiress, Marian Hooper, whose suicide in 1885 probably contributed to his general pessimism, and who is considered the model for his novel, *Esther* (1884, by "Frances Snow Compton"). His other novel, *Democracy* (anonymously published, 1880), investigates Washington politics. By 1877 he had left Harvard for Washington, and after his wife's death he travelled in the Orient, then returned to complete his *History of the United States during the Administrations of Jefferson and Madison* (9 vols., 1889-91). In his later years he travelled extensively from his Washington home, and was a president of the American Historical Association. He died at eighty, and was buried beside his wife beneath the St. Gaudens statue he had commissioned for her grave. *The Education of Henry Adams* (privately printed 1907, published 1918) complements his Mont-Saint-Michel in expressing his thermodynamic theory of history, the dynamo in his contemporary "multiverse" corresponding to the Virgin, who *unified* medieval thought. Remembered now as a historian, Adams can not be defined within any discipline, for his stimulating and original mind will appeal to every intellectual reader.

ADAMS, John (1704-40), a New England clergyman whose conventional *Poems on Several Occasions* (1745), which he described as "Musick for the Ear, Landskip for the Eye, and rich Repast for the Highest Understanding," imitated Pope, paraphrased the Bible, and included translations from Horace.

ADAMS, Samuel Hopkins (1871-), novelist and journalist of the muckraking school. *Revelry* (1926) fictionalized the Harding administration, and *The Harvey Girls* (1942) is a novel about the Fred Harvey chain of restaurants.

ADAMS, William Taylor (1822-97), wrote adventure stories for boys under the pseudonym of Oliver Optic, rivalling Horatio Alger in popularity and exceeding him in production with more than 1,000 short stories and 115 novels.

ADE, George (1866-1944), portrayed rural characters of his native Indiana with a keen sense of dialect and satire in *Fables in Slang* (1899); *People You Know* (1903); and *Hand-Made Fables* (1920). He also wrote farcical musical comedies of student life, including *The College Widow* (1904).

Agrarians; see Southern Agrarians.

AIKEN, Conrad [Potter] (1889-), was a contemporary of T.S. Eliot and Van Wyck Brooks at Harvard, where he was president of the *Advocate* and class poet. After publishing nine volumes of poetry, dating from 1914, he won the Pulitzer Prize for his *Selected Poems* (1929). Eight additional volumes have appeared since. His stories have appeared in *Esquire, The New Yorker, Scribners*, and many other magazines, and "Silent Snow, Secret Snow" is an anthology favorite. He has written four novels. *King Coffin* (1935), concerning criminal neurosis, reflects the keen interest in psychology which permeates all of his writing. He has deliberately eschewed literary coteries and associations, yet there is a notable resemblance in his later poetry to Eliot, even though he retains his musical rhythms. His stream-of-consciousness prose suggests an influence from Joyce.

AIKEN, George L. (1830-1876), actor and playwright, is now remembered solely for his dramatization of Mrs. Stowe's *Uncle Tom's Cabin*, which played 100 nights after its opening (1852) in Troy, New York. A melodramatic bag of episodes with no dramatic structure, it has probably played to more audiences than any play ever written.

AKINS, Zoë (1886-), has written two novels and two volumes of verse, but is best known for her plays. Her dramatization of Edith Wharton's *The Old Maid* won the Pulitzer Prize in 1935.

ALCOTT, [Amos] Bronson (1799-1888), an unsuccessful peddler who turned to teaching, in 1839 was forced to close his Temple School in Boston because of unfavorable reaction to his advanced theories and liberal attitudes. Using the conversational method, he sought the harmonious development of the whole personality, and was hailed by the English in 1842 as "the American Pestalozzi." The persisting poverty of the Alcott family was not relieved by his experiment (with two Englishmen) in communal farming at "Fruitlands," nor by his lectures, and the family achieved security only with the publication in 1868 of Louisa May Alcott's *Little Women*. Alcott wrote mainly about education, but he is related to the literary tradition through his intense Transcendentalism (q.v.) and his association with the greater writers of that philosophy. Emerson said, "As pure intellect, I have never seen his equal," but also called him a "tedious arch-angel."

ALCOTT, Louisa May (1832-88), daughter of Bronson Alcott, was tutored by him and by Emerson and Thoreau. From infancy she toiled to help the impoverished Alcott family, both at home and in domestic service. From the age of sixteen she also wrote, but with practically no success until 1863 when she received $200 for "hospital sketches" of her work as a volunteer nurse in Georgetown, an experience which, as with Whitman, wrecked her health. Her first novel, *Moods* (1865), afforded her a trip to Europe, where she first enjoyed the leisure to reflect on her writing, and as a result produced *Little Women* (1868) which won her fame and fortune, and established the pattern of her later work.

ALDEN, Henry Mills (1836-1919), one of the deans of American magazine editors, who, in his efforts to make *Harper's* a "family magazine," emasculated Hardy's *Jude the Obscure*.

ALDRICH, Thomas Bailey (1836-1907), was born in Portsmouth, N.H., the "Rivermouth" of his popular novel, *The Story of a Bad Boy* (1870). After the publication of his poems (*The Bells*, 1855) he was made junior literary critic of the New York *Evening Mirror*, and thereafter held a variety of editorial positions until, in 1881, he succeeded William Dean Howells as editor of the *Atlantic Monthly*. Like Howells, he had risen to this pinnacle in the temple of the New England literary

gods without the benefit of college training, and also like Howells, his conservative optimism made him a suitable guardian of the "Brahma" tradition. "Marjorie Daw" (1873) is his best remembered short story. His poems have been compared in grace and urbanity to the work of Horace. Although none of his themes in the novel, story, or verse reflect any depth of perception, he was a superb craftsman.

ALGER, Horatio, Jr. (1834-99), was the son of an extremely Puritanical minister who permanently warped the boy's personality. Alger finally rebelled by fleeing his graduation exercises at the Harvard Divinity School to become a Bohemian in Paris (1856). After his return he was persuaded by the father to become ordained as a Unitarian minister (1864), but he quit his pulpit two years later and settled in New York, eventually as chaplain of the Newsboys Lodging House. While he dreamed of writing a great book, he turned out some 130 success novels for boys, and became perhaps the most popular writer of all time with his formula that rags inevitably lead to riches if the behavior-platitudes are observed. Yet he remained lonely, unhappy, and conscious of his failure to produce any work of value.

ALGREN, Nelson (1909-), grew up in Chicago, worked in New Orleans and in Texas, and served in the Army during World War II. His first novel, *Somebody in Boots* (1935), is about hitch-hiking young hoboes during the depression. He first won critical acclaim with his novel of Chicago backroom life, *The Man with the Golden Arm* (1949). *Never Come Morning* (1942) is an earlier Chicago novel and *The Neon Wilderness* (1947) collects his short stories. He writes about the dispossessed and the poor with realistic understanding, not as a political agitator.

ALLEN, Frederick Lewis (1890-1954), was, until recently, editor of *Harper's*. His studies of contemporary decades, *Only Yesterday* (1931) and *Since Yesterday* (1940), provided a unique perspective of the American scene while it was still too close to be settled in its place as a part of our history. *The Lords of Creation* (1935), concerning the expansion of American corporations in this century, is refreshingly objective. He often collaborated with his brilliant wife, Agnes Rogers.

ALLEN, [William] Hervey (1889-1949), was prevented by an injury suffered in athletics from completing his courses at Annapolis, and graduated from the University of Pittsburgh in his native city. His *Wampum and Old Gold* was selected for the Yale Series of Younger Poets publication in 1921. The autobiographical novel, *Toward the Flame* (1926), described his war service, in which he was seriously wounded. *Israfel* (1926) has become a standard biography of Poe. He is popularly known for the best-selling *Anthony Adverse* (1933) a lengthy romance of the Napoleonic period based on careful historical studies and the diaries of his grand-uncle. He continued to produce historical novels,

but preferred to think of himself as a poet.

ALLEN, James Lane (1849-1925), grew up in a Kentucky impoverished by war, earned his way through Transylvania College where he later took his M.A. degree, and after several teaching positions became professor of Latin and English at Bethany College. Meanwhile he contributed articles, stories and poems to *Harper's* and the *Atlantic Monthly*. In 1885 he gave up teaching for a writing career. *Flute and Violin* (1891), containing Kentucky romances, marked his first success, and *A Kentucky Cardinal* (the bird), published in 1894, was his most popular work. *A Summer in Arcady* (1896) treated simple farm life with a degree of realism, but the dominating theme of his work was a romantic nostalgia for pre-war Kentucky life, painted in local colors.

ALLIBONE, S[amuel] Austin (1816-89), an insurance company executive who composed the valuable work, *A Critical Dictionary of English Literature and British and American Authors* (1858-71) in three volumes.

ALLSTON, Washington (1779-1843), came closer to literature in his paintings (*Dead Man Revived by Touching the Bones of the Prophet Elisha*) than in his poetry (*The Sylphs of the Seasons,* 1813), and is chiefly remembered for his literary associations. He married, first, the sister of William Ellery Channing and, after her death, the sister of R.H. Dana. During his long residence in England he was a friend of Southey, Wordsworth, and Coleridge, and he painted the portrait of the latter. He was also intimately acquainted with S.F.B. Morse, who later invented the telegraph.

American Academy of Arts and Letters, a division of the National Institute of Arts and Letters, founded in 1904 to promote American art, music, and literature and to honor the artists. Henry James, W.D. Howells, Henry Adams, and Mark Twain were early members.

American Folk-Lore Society, founded in 1888 to study and rescue from oblivion the folk literature of the Negroes, the American Indian, and the laborer in those industries which have stimulated an oral tradition, including transportation, the western cattle industry, and logging. Folk-lore is becoming an increasingly important part of the academic study of literature in America, and the research has been accelerated by modern inventions for sound recording. The Library of Congress has sponsored expeditions for the collection of material.

American Literature (1928-), the quarterly journal published by the Duke University Press in cooperation with the Modern Language Association and presenting valuable bibliographies in addition to critical studies of American literature.

American Mercury, The (1924-), was founded as a successor to the *Smart Set* by H.L. Mencken and George Jean Nathan with the avowed intention of seeking manuscripts which had been rejected by other editors, and an editorial policy of presenting the American mass mind in all of its glory or stupidity, but most frequently in the latter. After 1934, when both Mencken and Nathan had left the magazine, it became more concerned with economics or politics than with literature. Previously, it published most of the major contemporary American writers, including Dreiser, Sandburg, Ferril, O'Neill, and Sherwood Anderson.

American Prefaces (1933-43) was published by the University of Iowa under the guidance of Paul Engle, and published poems, stories, or articles by T.S. Eliot, Robert Frost, and Wallace Stegner, as well as by unknown writers.

American Spectator, The (1932-7), was founded by Nathan, Boyd, Sherwood Anderson, Cabell, O'Neill and Dreiser to continue the tradition of the *Smart Set* and *The American Mercury*, and demanding "clarity, vigor, and humor" of its contributors. In 1935 the editors became bored with the project, and left to less invigorating editors, it failed.

American Speech (1925-), a quarterly magazine founded and edited for eight years by Louise Pound, and devoted to the study of dialects and other specialties of the American language. It is presently edited by Allan Hubbell.

ANDERSON, Maxwell (1888-), the son of a Baptist minister, graduated from the University of North Dakota, and taught school for 2 years before going to Stanford, where he received an M.A.

degree in 1914. After teaching at Stanford and at Whittier College, he turned to journalism as being more remunerative, and in 1918 moved to New York where he wrote his first play, *White Desert*. With the success of *What Price Glory?* (1924), written in collaboration with Laurence Stallings, he became a professional dramatist. *Both Your Houses* won the Pulitzer Prize in 1933. He won the Drama Critics' Circle awards for *Winterset* in 1936 and for *High Tor* in 1937, perhaps his two best plays on account of the successful fusion of his social and poetic interests. Many of his plays are written in a loose blank verse, including *Elizabeth the Queen* (1930), *Night Over Taos* (1932) and *Mary of Scotland* (1933). In addition to a book of verse and radio dramas, he has produced more than 24 plays, either historical or treating an immediately contemporary issue, and they are almost always a financial success.

ANDERSON, Sherwood (1876 - 1941), from the age of 14 earned his living in Ohio cabbage fields, racing stables, and factories. The Spanish-American War broke this routine, and on his return from military service he entered the advertising business in Chicago. By this time he had become interested in creative writing, but it remained a latent interest for two decades. In his midthirties, married, and a successful manager of an Ohio paint factory, his conviction that he was "stuck in a rut" provoked him one day to leave both his business and his family, deliberately, abruptly, and permanently, and he never

ceased to be proud of this dramatic declaration of his "conversion." Back in advertising in Chicago, he met Sandburg, Dreiser, Ben Hecht, and Floyd Dell, and they encouraged him to publish his novel, *Windy McPherson's Son* (1916), concerning a boy's life in Iowa. This was followed by *Marching Men* (1917), chronicling the oppression of Pennsylvania coal miners. He also published a volume of poems, *Mid-American Chants* (1918). In 1919 his best-known work, *Winesburg, Ohio,* appeared, and it changed not only the personal history of the author but the course of American fiction. Such diverse writers as Hemingway, Faulkner, Wolfe, McCullers, and Saroyan (qq.v.) have been influenced by him, and the diversity as well as the superiority of these literary descendants indicates the beneficence of his subtle influence, which suggests but never dominates. Although the theme of *Winesburg,* as with most of his work, is a Naturalistic perception of the standardization and repression of human values by the machine age, its rare quality did not derive from intellectual perception. *Winesburg* was something that happened to Anderson, just as the break with the factory was something that happened to him, and all of his writing became a reiteration of that earlier conversion from middle-class mores. In his preface he described the characters of *Winesburg* as "grotesques" by reason of their embracing a single truth rather than the whole truth. They grope for life, but find only frustration. Anderson was not unique in perceiving this generalization about people. It was his sympathy and love for his characters that saved him from the sterility of mere documentation and made him unique as a writer. *Poor White* (1920) portrayed the destruction of the soul of a town by the arrival of the machine. *Dark Laughter* (1925) contrasted Negro laughter with White spiritual sterility in a simple but masterful narrative for which Anderson acknowledged his debt to Gertrude Stein, although she had declared him to be unmalleable when he was a member of her post-war Parisian cluster that included Hemingway, Pound, and Ford Madox Ford. After living in Paris, he spent a year in New Orleans, where he met and encouraged Faulkner. In 1925 he moved to a farm near Marion, Virginia and there edited both the Democratic and the Republican newspapers of the town. In 1941, while on a mission of goodwill, he died in Panama from complications that resulted from the sliver of a toothpick which a Washington hostess had used for her hors d'oeuvres.

ARP, Bill (1826-1903), the pseudonym of a Georgia lawyer, Charles Henry Smith, who wrote a series of letters for the Atlanta *Constitution* addressed to "Mr. Abe Linkhorn," and counterfeiting, in humorous southern dialect, a sympathy so illogical that the Northern cause was sharply satirized. The letters were later published in a book and were popular for years after the Civil War. Later in the war and during reconstruction the letters turned to rustic philosophy. Smith served as a major with the Confederate Army, later was a state senator, and in 1868 became mayor of Rome. His work includes *Bill Arp, So-Called* (1866), *Bill Arp's Letters* (1868), and *Bill Arp: From the Uncivil War to Date* (1903).

ARRINGTON, Alfred W. (1810-67), a minister, who became a lawyer in Missouri and Arkansas, a journalist in New York, and a judge in Texas. *The Desperadoes of the Southwest* (1847) gives his vivid account of the lynch law. *The Rangers and Regulators of Tanaha* (1856), a novel, depicts the Southwest in a transitional stage. His poems were published in 1869. He used the pseudonym, Charles Summerfield, in publishing his prose books.

ASCH, Sholem (1880-   ), a Polish writer whose novels, stories, plays and prose books depict the sufferings of the Jews in the modern world, both on the continent and in America. He became a resident of New York City in 1910, but since then has spent several years in Berlin and Paris. Most of his work has been translated from the German or Yiddish in which it was originally composed. *Three Cities, A Trilogy* (1933) recounts the history of St. Petersburg after 1905, Warsaw before 1914, and Moscow at the period of the revolution. Another trilogy concerns Jesus, *The Nazarene* (1939), portrayed as the greatest of the Jewish prophets; St. Paul, *The Apostle* (1943); and *Mary* (1949).

ATHERTON, Gertrude [Franklin] (1857-1948), wrote a series of novels using the history of California for background. She also wrote the biography of Alexander

Hamilton in the form of a romantic novel in which Hamilton was the unqualified hero, *The Conqueror* (1902). Her novel of the American girl who marries into European nobility, *Black Oxen* (1923), has the interesting episode of a 58 year old countess being transformed to the appearance and personality of 28 through a glandular operation. Mrs. Atherton was a great-grandniece of Benjamin Franklin, and a member of one of the influential pioneer families of California.

Atlantic Monthly (1857-) was named by Holmes, and Lowell was its first editor. Howells (editor 1871-1881) and his successor, Aldrich (editor 1881-1890), although apprentice Bostonians, retained the New England tradition while encouraging a few contributors from the outside. In 1871 the magazine paid Bret Harte $10,000 for his entire production of the coming year, to its subsequent regret. Many of Henry James's novels were serialized in the *Atlantic*. At the turn of the century it became more interested in the social scene, including muckraking. The *Atlantic* today retains its conservative tradition, resisting many of the temporary fashions in both literature and sociology but never quite deserving the label, reactionary.

AUDUBON, John James (1785-1851), was probably born in Haiti, the illegitimate son of a French naval officer, although he named New Orleans as his birthplace. He was educated in France and studied painting under David. His mercantile enterprises in Kentucky met with failure, but he was able to pursue his major interest in drawing the birds of the region. After years of poverty, supported by the labors of his faithful wife, he achieved recognition in Edinburgh in 1826, and his *The Birds of America* was published (1827-38) with an accompanying text by William Mac-Gillivray. His own ornithological writing was noted more for its enthusiasm than its accuracy. Many of his bird paintings with his notes were published over a period of 87 years after his death.

AUSLANDER, Joseph (1897-), a contemporary poet of the romantic tradition: *Sunrise Trumpets* (1924), *Cyclop's Eye* (1926), *Riders at the Gate* (1938) and *The Unconquerable* (1943) are a few of his several volumes. With Frank Ernest Hill he edited *The Winged Horse Anthology* of poetry.

AUSTIN, Mary [Hunter] (1868-1934), left her Illinois home for California at twenty. She lived in the lowlands of the Sierras and, after her marriage to Stafford W. Austin, on the Mojave Desert where she observed Indian life, described in *The Land of Little Rain* (1903). With Jack London and George Sterling she founded the literary colony at Carmel. Later she moved to Santa Fe to study the Indian and Spanish cultures of New Mexico. Her many interests are reflected in the variety of her books: *Isidro* (1905) is a historical romance set in California; *Santa Lucia: A Common Story* (1908) deals with the marriage problems of three city women; *A Woman* *of Genius* (1912) reflects her interest in the stage and in feminism. *The Ford* (1917) investigates social injustice, and *No. 26 Jayne Street* (1920) concerns radical elements of New York City. Her many other novels, stories, and plays are concerned with the danger of the machine to the individual, with the problems of women, or with the Southwest. Her personal research into the nature and processes of *Everyman's Genius* (1925) may be highly recommended to aspirant authors.

B

BABBITT, Irving (1865-1933), was born in Ohio, and educated at Harvard (M.A. degree 1893) and at the Sorbonne. He was a professor of French literature at Harvard from 1912 until his death. He was a leader in the Humanist movement (q.v.), a philosophy of both ethical and literary behavior. He opposed romanticism, which he felt had replaced the religious and classical with a decadent humanitarianism and utilitarianism. In strictly literary terms, his humanism now appears to stand about midway between the 19th-Century Romanticism and the various approaches to Realism in the 20th century. Among the works in which he expounded his doctrine are: *The New Laokoön* (1910) and *Rousseau and Romanticism* (1919).

BACON, Delia Salter (1811-59), the daughter of a Congregationalist minister, became a Baconian who thought that Raleigh and Spenser as well as Bacon wrote Shakespeare's plays, and had concealed in them, through ciphers, a whole system of philosophy. She temporarily converted Emerson to her theory, who sent her to Carlyle, but he rejected her when she rejected his suggestion that she consult the scholarship more and her

intuition less. Her insanity became manifest during the last two years of her life.

BAGBY, George William (1828-83), a Virginian who left the medical profession to become a journalist, is remembered as an early student of Negro dialect and as the author of humorous local-color sketches, which were selected for publication by T.N. Page in 1910.

BAILEY, James Montgomery (1841-94), is often considered to have been the first newspaper columnist and humorist, and was known as the "Danbury (Conn.) News Man." Throughout the Civil War he sent humorous sketches of army life back to his local newspaper, and later, as co-owner of the Danbury *Times,* his wit was largely instrumental in raising the circulation from two to thirty thousand in nine months, although an examination of his work leaves the modern reader quite sober. When even his own audience began to change in taste, he had the intelligence to maintain his circulation through sound and conservative journalism.

BAKER, Benjamin A. (1818-90), was an actor, theatrical agent, and manager before he wrote the play, *A Glance at New York* (1848), which mixed a new genre of realism with the standard melodrama in portraying Mose, a volunteer fireman and typical "Bowery Boy."

BAKER, George Pierce (1866-1935), who conducted his "47 Workshop" at both Harvard and Yale, achieved a phenomenal success in view of the number of his students who became successful dramatists, or, if they failed to achieve distinction in the theatre, were successful in other branches of literature. His students include Eugene O'Neill, Philip Barry, S.N. Behrman, Sidney Howard, Albert Maltz, John Dos Passos, and Thomas Wolfe. He is Professor Hatcher in Wolfe's *Of Time and the River*. His *Dramatic Technique* (1919) is a standard textbook in playwriting classes.

BALESTIER, [Charles] Wolcott (1861-91), is chiefly remembered for his collaboration with Kipling, his brother-in-law, on *The Naulahka* (1892), for which he supplied the American chapters. His *Benefits Forgot* (1892) is a novel set among Colorado mining camps.

BARKER, James Nelson (1784-1858), an early American dramatist whose *Superstition* (1824) was perhaps the best American play produced up to that time. Its subject was Puritanism in New England. His play about Pocahontas, *The Indian Princess* (1808) was the first American play with an Indian subject.

BARLOW, Joel (1754-1812), one of the "Hartford Wits" and a major contributor to their *Anarchiad,* had an amazing career as a soldier, teacher, publisher, businessman, poet, and diplomat, but is remembered for his charming little poem, *Hasty Pudding* (1796), and in spite of his tedious *The Columbiad* (1807), which he intended to be a great American epic. He published Paine's *Age of Reason,* backed Fulton's steamboat, effected treaties with Algiers to free American prisoners, made a fortune in France, and finally, as minister to France, died in Poland from efforts to consummate a treaty with Napoleon, whom he found in the disastrous retreat from Moscow.

BARRY, Philip (1896-   ), first achieved success as a dramatist with *You and I,* which won the Harvard Prize in 1922 and production on Broadway in 1923, followed by other comedies of manners. He is a master of witty dialogue and comic character contrasts, and frequently he shows a serious interest in the contrast between the commercial and the sensitive mind, as in *The Youngest* (1924), and *Holiday* (1928). *Hotel Universe* (1930) portrays an old man's mystical solution of many psychological problems facing his young guests. *Tomorrow and Tomorrow* (1931) and *The Animal Kingdom* (1932) are psychological investigations of contemporary marriage problems. *The Philadelphia Story* (1939) shows Barry at the height of his power as a wit and master of the drawing-room comedy.

BARTLETT, John (1820-1905), rose from employee at the University Book Store in Cambridge to senior partner at Little,

Brown & Co. With only a public school education, he wrote the most detailed Shakespeare concordance in existence (1894). Harvard awarded him an honorary M.A. His *Familiar Quotations* ran through nine editions before his death, and has since been republished 3 times. For years he was the unofficial bibliographer for Harvard University, both students and professors coming to him for bibliographical advice.

BATES, Katherine Lee (1859-1929), a professor at Wellesley who wrote "America the Beautiful" (1893).

BAUM, L[yman] Frank (1856-1919), journalist and playwright whose recently cinematized fantasy, *The Wonderful Wizard of Oz* (1900) was one of his 14 juvenile stories about that mythical land.

BAY PSALM BOOK (1640) is generally considered to be the first book published in America. It was the official hymnal of the Massachusetts Bay Colony.

BEACH, Joseph Warren (1880-), was long a professor of English at the University of Minnesota, and presently lectures at Johns Hopkins. As a critic he has specialized in American literature. *The Method of Henry James* (1918), *The Twentieth-Century Novel* (1932), and *American Fiction* 1920-1940 (1941) are his best known books. Beach is a gentleman scholar of the 19th-century pattern of English professor, but his mind is a match for any of the harsher moderns.

BEHRMAN, S[amuel] N[athaniel] (1893-), graduated from Harvard where he took the 47 Workshop of George Pierce Baker (q.v.). Later he worked toward the M.A. degree at Columbia under Brander Mathews and John Erskine. His first success, *The Second Man* (1927), is used by Joseph Wood Krutch as an example of the comic spirit in its purest form. His other comedies of manners include *Biography* (1932), *Amphitryon 38* (1937), and *No Time for Comedy* (1939). He has also written for Hollywood and for magazines.

BELASCO, David (1854?-1931), was born in San Francisco where, after a roving boyhood that included some years in Texas, he became locally prominent as actor, playwright, and producer. Later he moved to New York where he was intimately associated with the theater for half a century, and achieved fame in almost every department of play production, developing new actors, inventing new staging devices, writing, managing, and producing. Among his outstanding successes were *Madame Butterfly* (1900), written with J.L. Long, and *The Girl of the Golden West* (1905), later written as an opera by Puccini. M.J. Moses edited an edition of six of Belasco's plays in 1928.

BELL, James Madison (1826-1902), a Negro poet and radical abolitionist, aided his friend, John Brown, in recruiting men for the 1859 raid. After the war he entered politics and campaigned for Grant. *The Poetical Works of James Madison Bell*, Byronic in tone, were published in 1901.

BELLAMY, Edward (1850-98), a Massachusetts novelist and journalist, first became interested in economic inequalities during a trip to Europe in 1868. He edited the Springfield *Union,* and in 1880, with his brother, founded the Springfield *Daily News.* Meanwhile he contributed stories to magazines, and in 1879 his first novel, *The Duke of Stockbridge,* was serialized, although he failed to finish the story. Other novels, concerning psychic phenomena, followed, but he is remembered chiefly for his provocative *Looking Backward* (1888), which sold more than a million copies and stimulated the formation of Bellamy clubs and the *Nationalist Party.* It pictures Julian West waking from a hypnotic sleep in the year 2000 to find America has become a socialist utopia. With materialism controled, cultural maturity was achieved. *Equality* (1897), its sequel, is merely a political tract, although his determination to finish it before seeking alleviation of his tuberculosis in Colorado probaby hastened his early death.

BEMELMANS, Ludwig (1898-), was exiled to America in 1914 by his Austrian family after a prolonged record of disobedience at schools and as a worker in his uncle's resort hotels. In New York he studied art while advancing in the hotel business until he became proprietor of his own restaurant. *My War with the United States* (1937) is a humorous account of his Army service. Most of his other books relate experiences in his business or describe foreign travel, always with his unique and whimsical

wit, and illustrated with his charming drawings. Among these are *Hotel Splendide* (1941), *I Love You, I Love You, I Love You* (1942), and *Hotel Bemelmans* (1946), a collection of his stories about hotel life in New York City.

BENÉT, Stephen Vincent (1898-1943), grew up at army posts in California and Georgia, took his B.A. and M.A. degrees at Yale, and also studied at the Sorbonne. His long narrative poem, *John Brown's Body* (1928), and *Western Star* (1943) both won Pulitzer Prizes, and his *King David* took the *Nation's* poetry prize in 1923. In 1932 he received the Shelley Memorial Award, and in 1933 the Roosevelt Medal. In addition to many volumes of poetry, he wrote five novels, the librettos for two one-act operas (*The Headless Horseman,* 1937, and *The Devil and Daniel Webster,* 1939), and short stories. Most of his work is concerned with folk or epic aspects of the American scene, which he handles with great technical facility whether using the ballad form, or experimenting with a variety of forms as he did in *John Brown's Body.* His earlier work has a jaunty wholesomeness that was modified in some of his later work by a sharper understanding of historical forces.

BENÉT, William Rose (1886-1950), brother of Stephen Vincent Benet and husband of Elinor Wylie, was a poet, editor, and anthologist. His autobiographical verse novel, *The Dust Which Is God,* won the Pulitzer Prize for 1942.

BENSON, Sally (1900-), writes short stories mainly for *The New Yorker,* which have been collected for publication in book form. *Junior Miss* (1941) and *Meet Me in St. Louis* (1942) were successfully dramatized. In her rather quiet stories she often dramatizes profound human tragedies by means of trivial domestic incidents; in "The Seeing Eye," a dominating wife tells her blind husband that he is dragging his sleeve in the butter, a prevarication engineered to break his conversation with a more attractive woman.

BERCOVICI, Konrad (1882-), a Rumanian musician, came to New York from Paris in 1916 to make a concert tour, where a skating accident ended that career. While recuperating he began his writing career. His many short stories and novels deal with Gypsy life in Europe and America, or more frequently with the problems of the immigrant. *Dust of New York* (1919) presents sketches of the New York foreign neighborhoods. His novels include *The Volga Boatman* (1926) and *Exodus* (1947).

BERNARD, William Bayle (1807-1875), wrote more than a hundred successful plays, including a dramatization of *Rip Van Winkle* (1832). He was perhaps the first to satirize the "country rube" for the American stage.

BEVINGTON, Helen (1906-), writer of light verse, most of it appearing in *The New Yorker*, was born in upstate New York. Since 1942 she has lived in Durham, North Carolina, where both she and her husband teach English at Duke University. Frequently her subjects concern literary personages, many

of them from 18th-century England. Vivacity and grace characterize the seasonal and *locale* poems. *Doctor Johnson's Waterfall* (1946) and *Nineteen Million Elephants* (1950) are her two collections.

BIERCE, Ambrose (1842-1914?), escaped from his hated, impoverished Ohio family by enlisting as a drummer boy in the Civil War, and rose to brevet major. Later he became a successful journalist in San Francisco. While in England (1872-76) he contributed to several journals and published three books, including *Cobwebs from an Empty Skull* (1874). Back in San Francisco, his brilliant and bitter wit gave him a veritable dictatorship over California literary circles. But the nineties were not good to Bierce: his wife left him ('91), one son was killed in a brawl ('89), the other son died an alcoholic (1901), and the changing times dissipated his popularity. In 1898 he went to Washington and for ten years was correspondent for the New York *American*. His objective bitterness turned increasingly upon himself, and in 1913 he again escaped into a civil war, this time the Mexican. Nothing is known of him from the day he crossed the border in 1913. It was during his later, especially unhappy period that he produced his best work. *Tales of Soldiers and Civilians* (1891) and *Can Such Things Be?* (1893) have the Poe horror delineated with Stephen Crane realism, alleviated or sometimes intensified by Mark Twain humor, and handled with O. Henry dexterity. With so much to carry, sometimes his stories fall flat of their own weight. ("Oil of Dog" portrays a woman engaged in the disposal of unwanted babies and her husband operating a dog oil factory. They decide to join operations when they accidentally discover the superiority of baby to dog oil.) *The Devil's Dictionary* (1911) of ironic definitions lists the barometer as an ingenious instrument which tells what kind of weather we are having; "discussion" is a method of confirming others in their errors; "happiness" is contemplation of the misery of another. Each of these definitions, to some degree, may be said to define Bierce.

BILLINGS, Josh (1818-85), the pseudonym of Henry Wheeler Shaw, who failed at a variety of occupations, both at his Massachusetts home and on the western frontier, until, at the age of 45, he developed the personality of a homely crackerbox philosopher that became popular in newspapers, on the lecture platform, and in books: *Josh Billings, His Sayings* (1865), *Josh Billings on Ice* (1868), *Farmer's Allminax* (issued annually, 1869-79) and four others. He exploited the devices of the mid-century newspaper humorist: eye-dialect, inverted logic, ridiculous grammar, monstrous spellings, and word distortion. Yet, there was a wisdom and social satire that raised him above the average word-juggler: "I argue this way—if a man is right, he can't be too radical. If he is wrong, he can't be too conservative."

BIRD, Robert Montgomery (1806-54), received an M. D. degree from the University of Pennsylvania, but practised only one year, then turned to playwriting. *The Gladiator* (1831) was the first play written in the English language to be performed more than a thousand times during the dramatist's life, and became a favorite role of the tragedian, Edwin Forest. In addition to plays, Bird wrote novels based on Mexican or American history, his best being *Nick of the Woods* (1837), which presents realistic Indians to contradict Cooper's "noble savage."

BISHOP, John Peale (1892-1944), southern poet, novelist, and short story writer won popularity almost exclusively with the academic critics. A classmate of Fitzgerald at Princeton, he is said to have been the prototype of Tom D'Invilliers in *This Side of Paradise*. His poetry collections include *Green Fruit* (1917) and *Minute Particulars* (1935). *Act of Darkness* (1935) is a novel.

BLACKMUR, Richard P. (1904-), poet (*From Jordan's Delight*, 1937; *Second World*, 1942), teacher, and critic, was born at Springfield, Mass. He was an editor of *Hound & Horn* from 1927 to 1934. He is best known as a critic: *The Double Agent* (1935), *The Expense of Greatness* (1940), and *Henry James* (1941). Although he has been identified with the "New Critics," he is less of an absolutist than most of the members of that movement.

BLY, Nelly, pseudonym of Elizabeth Cochrane Seaman (1867-1922), which she took from Stephen Foster's song. A crusading reporter, she first exposed mining conditions for the Pittsburgh *Dispatch,* and later, reporting for Pulitzer's New York *World,* feigned insanity to get her material for *Ten Days in a Mad House* (1887). *Nelly Bly's Book: Around the World in Seventy-two Days* (1890) was based on the trip she took under sponsorship of the *World,* which brought her international fame.

BODENHEIM, Maxwell (1892-1954), born in Mississippi, became one of the Imagist poets of the twenties, and later degenerated into a Greenwich Village "character," impoverished and impotent. In addition to nine volumes of poetry, he wrote several novels and plays.

BOGAN, Louise (1897-), a *New Yorker* critic, was born in Maine, and attended Mount St. Mary's Academy and Boston University. In addition to four books of poetry, she wrote *Achievement in American Poetry* (1951), a history of the poetry "schools" since 1900. As a critic, she parallels F. R. Leavis in her partiality for T. S. Eliot.

BOKER, George Henry (1823-90), dramatist son of a wealthy Philadelphia banker, is best remembered for his verse tragedy, *Francesca da Rimini* (1855). He also wrote forceful sonnets on public affairs and love sonnets that were published by his biographer, E. S. Bradley, in 1929. He served as minister to both Turkey and Russia, and was instrumental in clearing his father of the false charge that he had wrecked the Girard Bank.

BONER, John Henry (1845-1903), lyric poet much admired in his day, was espe-

cially gifted in the expression of nostalgic moods. Born in Salem, North Carolina, he aligned himself with the Republican party during Reconstruction and thereafter found it best to leave his state for a political assignment in Washington, D.C. His dainty, somber verses, many of which reveal his upbringing in the Moravian church, are collected in *Whispering Pines* (1883) and *Poems* (1903).

BOYD, James (1888-1944), historical novelist, was born in Pennsylvania, attended Princeton and Cambridge, saw service in France during World War I, then retired to Southern Pines, North Carolina. *Drums* (1925) is a novel of the Revolution in tidewater North Carolina, and *Marching On* (1927) deals with the Civil War in the same general area. *Long Hunt* (1930) is set along the Tennessee frontier. *Roll River* (1935) is a story with a Pennsylvania setting, and *Bitter Creek* (1939) is more western than historical. All these books disdain romantic trappings and attempt to give a true picture of the times. *Eighteen Poems* (1944) and *Old Pines and Other Stories* (1952) were published posthumously.

BOURNE, Randolph Silliman (1886-1918), was born in Bloomfield, New Jersey, and suffered throughout his short life from physical deformity. While a student at Columbia (from which he graduated in 1913) he became a leader in the early criticism of American institutions from a literary point of view, and of literature from an institutional point of view. In his books, *Youth and Life*

(1913), *The Gary Schools* (1916), *Education and Living* (1917), and *Untimely Papers* (posthumously collected, 1919), he criticized sentimentality in literature, the mores of big-business, and rationalizations in behalf of war. Van Wyck Brooks described his career in *The History of a Literary Radical* (1920).

BOYLE, Kay (1903-   ), was fortunate in having a mother who encouraged her to write from early childhood. Four volumes of her impressionistic short stories have been published, including *The White Horses of Vienna* (1936) and *The Crazy Hunter* (1940). Two of her stories have won the O. Henry Memorial prize award. Her novels usually concern Americans in Europe, where she lived for many years. Her later works picture the Nazi movement, leading up to and including the war. In *Death of a Man* (1936), an American wife of an Englishman rejects the love of a Nazi doctor. *Primer for Combat* (1942) describes the occupation of France, and *Avalanche* (1943) the resistance. In her total work she has covered a great variety of people, scenes, and economic classes.

BRACKENRIDGE, Hugh Henry (1748-1816), was born in Scotland, and brought to a Pennsylvania farm as a boy, where he was educated by his mother. He worked his way through Princeton, his first school, by acting as master of the grammar school, and there became intimate with Madison and Freneau. Receiving a master's degree in theology, he served as a chaplain through the Revolutionary War, but religious doubts later caused him to shift to the legal profession, and in 1781 he settled in the frontier village of Pittsburgh. Here he engaged actively in politics, considerably hampered by his ability to see both sides of a question, and his satire concerning the premium which democracy places on ignorance and stupidity, *Modern Chivalry* (1792-1815) is still both entertaining and provocative. His verse dramas, however, were stilted and pompous.

BRADFORD, Gamaliel (1863-1932), a biographer of American historical figures (*Lee, the American,* 1912; *Union Portraits,* 1916; *Damaged Souls,* 1923; etc.), he used a system which he called "psychography" to synthesize out of all the biographical data available a psychological portrait of the man. He is thus of historical importance to the study of literary criticism, standing as a precursor to the psychoanalytical critics such as Van Wyck Brooks.

BRADFORD, Roark (1896-   ), has written several interpretations of Negro life. *Ol' Man Adam an' His Chillun* (1928) was dramatized by Marc Connelly as *The Green Pastures.*

BRADSTREET, Anne [Dudley] (c. 1612-72), was reared in a cultural and liberal environment, although her father was a Puritan and steward to the Earl of Lincoln, also a Puritan. At 16 she was married and at 18 came to New England with her father and husband, both of whom later became governors of the Massachusetts Bay colony. Despising the frontier environment, perhaps she turned to poetry to sublimate the initial misery, although colony life was considerably improved by the time her poems were taken to London by an admiring brother-in-law and in 1650 published as *The Tenth Muse Lately Sprung Up in America*. Her longer poems, deriving from Spenser, Sidney, Raleigh, and DuBartas, were feeble and narrow. Her more intimate, shorter poems are pleasant, but it is improbable that they would have justified publication had she not aroused curiosity by virtue of her being the first female poet of America.

Brahmins: a name derived from the highest Hindu class and applied to Boston society. In literature the title is used to define the entrenched literary conservatism which held to the mid-nineteenth century attitudes, represented by Lowell, Holmes, Emerson, and Longfellow, long after their vitality and pertinence had passed away. The mores of New England social behavior and the test of "good taste" were mixed with purely literary considerations as a dichotomy grew up between New England and the fluid frontier, representatives of two extremes.

BROMFIELD, Louis (1896-   ), is a native of Ohio, where he still operates an experimental farm. After two years at Cornell and Columbia, he joined the French army, and won the *Croix de guerre*. Later he was made a chevalier of the French Legion of Honor. He lived for many years in France, always preferring the country to resort or metropolitan centers. His first novel, *The Green Bay Tree* (1924), examines an American

steel town in social transition, *Early Autumn* (1926) won the Pulitzer Prize, and like much of his work is concerned with escape from tradition. *The Rains Came* (1937) and *Night in Bombay*, (1940) both reflect his knowledge of India, which he has visited frequently. Mr. Bromfield's subject matter ranges over half the world for its setting, and his characters may be dowagers in Italy or New York, or silver miners in Colorado.

Brook Farm (1841-7): a utopian experiment in communal living sponsored by the Transcendental Club of Boston, and established near West Roxbury, Massachusetts. It was the subject of Hawthorne's *The Blithedale Romance*. He was the only important Transcendentalist to live there, although Emerson, Channing, Alcott, and Margaret Fuller were interested in the project. It was proposed to secure the greatest cultural, physical, and moral education in a simple life where all labor and profit was shared. Although the farm was more successful and enduring than many similar experiments, a damaging fire dampened the enthusiasm of the members, and they dissolved after six years.

BROOKS, Maria Gowen (c. 1794-1845), became the ward of her brother-in-law at 15 and shortly thereafter married him, probably out of necessity. After an unhappy early life, she inherited a coffee plantation in Cuba, and lived there until her sons were ready for college. She spent a year in England among the "Lake Poets," and won undue praise from Southey, who had named her the "Maria of the West" upon the publication of her first volume, *Judith, Esther, and Other Poems* (1820). She also wrote *Zóphiël* (1833) and the novel, *Idomen* (1843), based on her unhappy love affair with a Canadian army officer. Some of her poetry is remarkable for the period, and had she published less of her inferior work she might have been better remembered.

BROOKS, Van Wyck (1886-), was born in Plainfield, New Jersey, graduated from Harvard, and did editorial work in England for the *Standard Dictionary*. While he was a temporary expatriate, he wrote *The Wine of the Puritans* (1909) which first presented his theory that the Puritan tradition had emphasized materialism at the expense of our aesthetics. This premise later became the thesis of his psychoanalytical criticism of *The Ordeal of Mark Twain* (1920) and *The Pilgrimage of Henry James* (1925), in which he used the men to argue his theory. His later work presents a more objective analysis of American literary history. *The Flowering of New England*, from 1815 to 1865, won the Pulitzer Prize in 1937, and he won honorary degrees that year from Columbia, Tufts, and Bowdoin. *New England: Indian Summer* (1940) continued the earlier history to 1915. *The World of Washington Irving* (1944) covered 1800-1840 outside New England, and the history is completed with *The Times of Melville and Whitman* (1947). *The Writer in America* (1953) is his latest study of American civilization.

BROWN, Charles Brockden (1771-1810), is remembered as America's first professional author. Had he possessed the quality of critical discipline, he might have been remembered as one of the great novelists. Unfortunately, an indulgent family persuaded him of his genius at an early age. He wrote in haste and seldom revised. But he was unique among the Gothic novelists in replacing the usual bag of tricks with a psychological motivation of the horror. *Wieland* (1798) presents a religious farmer who kills his wife and children at the command of a mysterious voice, which he accepts as the word of God. His other novels, *Arthur Mervyn* (1799)— portraying a yellow fever epidemic; *Ormond* (1799)— picturing murder and rape; the detective story of *Edgar Huntly* (1799); and *Clara Howard* (1801); suffer more than *Wieland* from a lack of plan, but show flashes of genius. He was admired by Keats, Shelley, Scott, and Godwin. Finding it impossible to sustain a family by professional writing, he engaged in trade and eked out a living by hack journalism on the side, while being constantly harried by ill health.

BROWN, John Mason (1900-), drama critic of the *Saturday Review of Literature*, was a member of George Pierce Baker's Drama Workshop at Harvard, where he later taught the history of the theater. He has studied the dramatic art and production methods of most European countries at first hand. His books include: *The Modern Theater in Revolt* (1929), *The Art of Playgoing* (1936) and *Many a Watchful Night* (1944). Mr. Brown is also one of the most dynamic public speakers in America.

BROWN, William Hill (1765-93), is presumed to have been the author of *The Power of Sympathy*, which in turn is presumed to be the first American novel. In 1789 a Boston printer issued the epistolary two-volume work anonymously, perhaps because Brown was a neighbor of the Mortons; a scandal in their immediate family was the basis of the story. In 1878 Francis Samuel Drake listed Mrs. Sarah Morton (q.v.) as the author—even though the Mortons had bought and destroyed all the copies they could find. In 1894 Brown's niece suggested that there were many reasons to suppose he had written the book, and her theory has since been substantiated by scholarship. Brown wrote other fiction, a series of essays, and many poems, including 24 uncollected verse fables. He also wrote two dramas: a comedy, *Penelope*, and a tragedy concerning Major André, *West Point Preserved*, which merited 27 performances at the Haymarket Theater in 1797. He died in Murfreesboro, North Carolina while visiting a sister.

BROWNE, Charles Farrar: See Ward, Artemus.

BROWNE, John Ross (1821-75), was a precursor to Mark Twain in satirizing the conventions of travel (c.f. *Yusef*, 1853, to *Innocents Abroad*) and in frontier humor (c.f. *Adventures in the Apache Country*, 1869, to *Roughing It*).

He also preceded Melville in writing *Etchings of a Whaling Cruise* (1846).

BROWNELL, W[illiam] C[rary] (1851-1928), paralleled Matthew Arnold in his literary criticism. Although he preceded the Humanists (q.v.) he is often identified with them, since he felt that a return to past culture would cure the United States of its provincialism. His books include *French Traits, An Essay in Comparative Criticism* (1889); *French Art* (1892); *Victorian Prose Masters* (1901); *American Prose Masters* (1909); *Criticism* (1914); *The Genius of Style* (1924); and *The Spirit of Society* (1927). See also, CRITICISM, Humanism.

BRYANT, William Cullen (1794-1878), son of a prominent physician and legislator, grew up in western Massachusetts where the primitive Hampshire Hills engendered his enduring love of nature, so prominent in his poetry. A precocious child, he showed phenomenal scholarship when he was sent to a clergyman uncle to study Latin, and then to another minister to learn Greek. At 13 he wrote an anti-Jefferson satiric poem, "Embargo," which was printed two years later. He wrote his great "Thanatopsis" and "To a Waterfowl" at 17, one of the most amazing achievements for a poet of that age in literary history. Six years later when his father presented them to the *North American Review*, without the poet's knowledge, Richard Henry Dana suspected a hoax, for the quality of the style far exceeded that of even the recognized American poets. When Dana published them, Bryant's reputation was firmly established.

Meanwhile, Bryant had been admitted to the bar, and practised law, which he abhorred, until 1825 in Plainfield, Massachusetts. In 1824 he contracted to supply the *United States Literary Gazette* with a hundred lines of verse a month for payment of $200 a year. Editorial positions followed, and in 1829 he became editor of the New York *Evening Post,* in which position he remained for the rest of his life, gaining half ownership of the paper by 1840. He was successful as a journalist, and on one occasion had to return from one of his many trips to Europe in order to save the *Post* from the mismanagement of a radical deputy.

After his early success, Bryant became established as America's leading poet, or, as he was often designated, the "American Wordsworth." His 1832 collection of *Poems* contains his major and definitive work, although other volumes were issued thereafter. The poet of nature, who had achieved an impeccable style at 17 and had thereafter diverted his intellectual perceptions to journalism, had small material for development and little room for improvement.

BUCK, Pearl [Sydenstricker] (1892-), the daugther of missionaries, spent her early years in China, and after attending school in America, she returned to teach at Chinese universities. Most of her novels have a Chinese setting, the best known being *The Good Earth* (Pulitzer Prize, 1932), which she wrote in three months. Mrs. Buck was awarded the Nobel Prize in 1938, the first American woman to be so honored.

BUNNER, H[enry] C[uyler] (1855-96), was, for most of his adult life, editor of the first humorous magazine in the United States, *Puck* (1878-96), to which he contributed his excellent light verse and parodies (as, "Home, Sweet Home" in the style of Whitman). His stories and sketches resembled Maupassant, and as a chronicler of Manhattan he was a precursor of O. Henry.

BURGESS, Gelett [Frank] (1866-1951), San Francisco humorist who wrote "The Purple Cow," a quatrain that has moved from the closet to the oral tradition. It was first published in *The Lark* (1895-7), a collection of the work of Les Jeunes, a California literary group.

BURKE, Kenneth [Duva] (1897-), was educated at Ohio State University and at Columbia, did editorial and research work in New York City, and was music critic for the *Dial,* and later for the *Nation,* where much of his literary criticism also appeared. At his best, *Counter-Statement* (1931), he is one of the greatest of contemporary critics, and also, while writing within the modern framework of ideas, one of the most unusual. Like Aristotle, he often uses the empirical inquiry concerning aesthetics: not, what should work, but what has worked? Thus he contrasts "the psychology of the audience," using Shakespeare for his example, with much of the contemporary literature which substitutes "the psychology of the hero."

Shakespeare created appetites and then satisfied them. His other books include *The Philosophy of Literary Form* (1941), and a collection of stories, *The White Oxen* (1942). See also CRITICISM.

BURNETT, W[illiam] R[iley] (1899-), is the author of *Little Caesar* (1929), a novel about Chicago gangsters which was a Literary Guild selection. Much of his other work follows this type of subject matter, dealing with gangsters, jazz performers, prizefighters, and sportsmen. *Goodbye to the Past* (1934), perhaps his best novel, begins with the protagonist's death and works back to his youth. Burnett was influenced by Prosper Mérimée, Pio Baroja, and Giovanni Verga in his emphasis on drama and condensation.

BURT, [Maxwell] Struthers (1882-1954), attended Princeton and later taught there, ranched in Wyoming, served with the Air Corps in World War I, and finally settled in North Carolina as a successful novelist and writer of short stories: *The Interpreter's House* (1924); *The Delectable Mountains* (1927); and *John O'May, and Other Stories* (1918).

BYNNER, [Harold] Witter (1881-), after graduation from Harvard in 1902 became assistant editor of *McClure's Magazine*, later taught at the University of California, and spent more than two years in the Orient collecting and studying Chinese literature and art. *The Jade Mountain* (1929), a translation with Dr. Kian Kang-hu of Chinese poetry from the T'ang dynasty, was a major service because of Bynner's ability to make the Chinese intelligible to American readers. Beginning with a melodious lyricism in his own poetry, *An Ode to Harvard* (1907), Bynner experimented with various approaches to the conservative tradition, the most successful resulting in *Eden Tree* (1931). His parody of contemporary verse, *Spectra* (1916, "by Emanuel Morgan and Anne Knish"), written in collaboration with Arthur Davidson Ficke, was a hoax so successful —or unsuccessful, depending upon the point of view—that it was taken seriously by many critics who might contend that it contained some of the best poetry Bynner ever wrote. He is now residing in Santa Fe, New Mexico.

BYRD, William (1674-1744), a Virginia planter, surveyor, bibliophile, and statesman, was educated in England, traveled on the continent, and was reputed to be as much at home at the English court as in the Virginia fields. He laid out the city of Richmond, on his own land, in 1733. His diaries, written as an avocation, were published posthumously, as manuscripts were discovered, from a hundred to two hundred years after his death: *The Westover Manuscripts* (1841), and *Another Secret Diary* (1943). Reporting on such matters as the survey of the boundary between North Carolina and Virginia, Byrd wrote with a keen wit and urbane charm that was far too uncommon among colonial reporters. Thackeray admired him and used his work as a reference in writing *The Virginians.*

C

CABELL, James Branch (1879-   ), is of an old Virginia family. He has been a journalist in New York and Richmond, and a coal miner in West Virginia (1911-13), but above all else he has been a novelist since *The Eagle's Shadow* was published 50 years ago. His exotic fantasies set in medieval "Poictesme" are written with a curious romanticism that he constantly satirizes—by innuendo, sarcasm, or irony. *Jurgen* (1919) is the best known of the many novels in this series. He has also written poems of an archaic flavor (*Sonnets from Antan,* 1929), criticism that protests the trend toward realism, short stories, and essays on his native state.

CABLE, George Washington (1844-1925), combines three disparate American traditions: his father was of old Virginian stock; his mother was rock-ribbed Puritan; and he supported his family from the age of fourteen. He fought on the Confederate side of the Civil War without enthusiasm, and while recuperating from this experience, began to write for the New Orleans *Picayune.* He was fired for refusing to review a play, which the tenets of his church would not allow. Probing the archives of New Orleans, he found the material for his stories that were published as *Old Creole Days* (1879). He then wrote a historical romance containing anti-slavery propaganda, *The Grandissimes* (1880), and the novelette about a quadroon, *Madame Delphine* (1881). His growing concern with social evils resulted next in *Dr. Sevier* (1885), exposing prison evils, and *The Silent South,* which spoke for the Negro. He was a leader of the local-color movement in the South, as well as a reformer. In 1885 he moved to Massachusetts, where he established the Home-Culture Clubs with a gift from his friend, Andrew Carnegie.

CAIN, James M[allahan], (1892-), graduated at 18 from Washington College, of which his father was president, and entered journalism in Baltimore, where H.L. Mencken encouraged him in his writing ambitions. *The Postman Always Rings Twice* (1934) is about a young poolroom hoodlum who murders the husband of the woman he loves. His other novels include *Serenade* (1937), *Past All Dishonor* (1946), and *The Moth* (1948).

CALDWELL, Erskine [Preston] (1903-), son of a Georgian minister, had no pre-

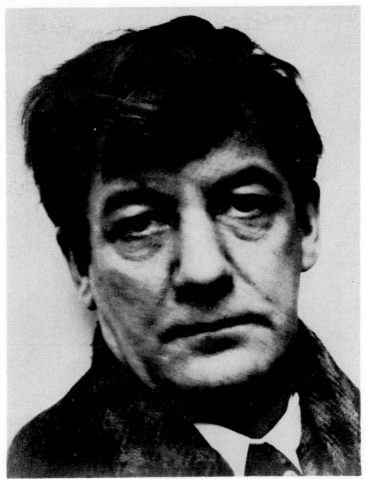

Photo: Alfred Stieglitz

SHERWOOD ANDERSON

MARY AUSTIN

JAMES BRANCH CABELL

Photo: Lotte Jacobi

ERSKINE CALDWELL

paratory schooling, but saw much of the raw life he later described in his novels, before he entered the University of Virginia. His best known social studies of southern tenant farmers, *Tobacco Road* (1932) and *God's Little Acre* (1933), unfortunately owe their fame to the publicity given them by societies for "the suppression of vice." Actually, the suppression of a particular vice was their motivation, a sociological purpose that Caldwell relieved with rich folk humor.

CALEF, Robert (1648-1719), a Boston merchant, bitterly attacked Cotton Mather and other Puritans responsible for the Salem witchcraft trials in *More Wonders of the Invisible World* (1700, London). The Mathers made no reply to his documentation, but attacked the book vigorously and, according to legend, Increase Mather had it burned in the Harvard Yard.

Calvinism: See "Puritanism."

CAMPBELL, Walter Stanley; see Vestal, Stanley.

CANBY, Henry Seidel (1878-), was one of the founders of *The Saturday Review of Literature,* and editor until 1936. He was also chairman of the board of judges of the Book-of-the-Month Club.

CANFIELD [Fisher], Dorothy (1879-), grew up in Kansas, where her father was a professor, took her A.B. degree at Ohio State where he was President, and received her Ph. D. degree at Columbia where he held the Chair of Librarian. Since her marriage she has lived in Vermont, except for some residence abroad. She has written novels, short stories, essays on education, criticism, books and plays for children, and translated Papini's *Life of Christ.* Her novels include *The Bent Twig* (1915), *Rough-Hewn* (1922), and *Deepening Stream* (1930).

CANNON, Legrand, Jr. (1899-), studied philosophy at Yale and Business Administration at Harvard (B.M.A., 1922). In 1932 he gave up business for writing historical novels about New Hampshire, where he spends much of his time. They include *A Mighty Fortress* (1937) and *Look to the Mountain* (1942). *The Kents* (1938) is about Wall Street financing in the late 19th century.

CAPOTE, Truman (1925-), grew up in and around New Orleans, and began writing and publishing stories before he was twenty. In 1948 he won the O. Henry Memorial Award for the best short story of the year. His stories have been collected in *Tree of Night* (1949). *Local Color* (1950) is a book of sketches of people and places. His first novel, *Other Voices, Other Rooms* (1948) pictures the growing pains of a 13-year-old boy who has homosexual tendencies. *The Grass Harp* (1951) pictures one sister escaping another sister to live in a tree, and was successfully produced as an experimental play at the Circle in the Square Theater in New York City.

CARMAN, [William] Bliss, (1861-1929), Canadian-born poet, rebelled against the gentle poetry of his day by introducing a gypsy quality into his poetry, sometimes pagan, often robust. *Songs from Vagabondia* (1894), written in collaboration with Richard Hovey, was followed by *More Songs* and *Last Songs* (1896 and 1901). He was fascinated in his later work with musical language and symbolic overtone: *Sappho* (1904), *Pipes of Pan* (1906), and several other volumes.

CATHER, Willa [Sibert] (1876-1947), was born in Virginia, but her family moved to a ranch near Red Cloud, Nebraska when she was nine, and she grew up among the foreign-born pioneers who were beginning the struggle to make farming pay in this area. She received her education at home until she started high school, and later graduated from the University of Nebraska. She was a journalist and teacher in Pennsylvania before joining the staff of *McClure's Magazine* in 1906. After the publication of her first novel, *Alexander's Bridge* (1912), she devoted her time exclusively to writing. Advised by Sarah Orne Jewett, whom she greatly admired, to write from her intimate knowledge of her own background, Miss Cather next produced one of the great novels of agricultural pioneering, *O Pioneers!* (1913), strong and stark in its portrayal of Alexandra Bergson's rural sophistication. *The Song of the Lark* (1915) takes a Colorado girl to success at the Metropolitan Opera. *My Antonia* (1918) and *One of Ours* (Pulitzer Prize, 1923) return to Nebraska farm scenes. She turned from a study of feminine strength to feminine charm in her sensitive portrait of *A Lost Lady* (1923), whose love for her husband's pioneering strength wanes when he, much older than she, retires among the drab sons of pioneers in Eastern Nebraska. *The Professor's House* (1925) breaks between the story of an unconventional scholar and a self-reliant student who personifies the professor's ideals. *Death Comes for the Archbishop* (1927) and *Shadows on the Rock* (1931) both go back in history to study the spiritual pioneering of the Catholic Church, the first in New Mexico, the second in Quebec. Miss Cather never achieved the technical proficiency of Henry James, who was her first model, and is at her best when technique may be restricted to the realistic manipulation of characters rather than events. Never relying on pyrotechnics or exotic settings, never descending to the mud flats of realism, never raising her voice against man or God, she had an uncanny ability to maintain a quiet intensity of interest. Her production of so many fine novels, quantity with quality, gives her some claim to rank as the greatest American novelist of the 20th century.

CHANNING, William Ellery (the younger) (1818-1901), nephew of the elder Channing, devoted his life to writing but never to polishing his poetry, published in five volumes from 1843 to 1886, and highly imitative of Emerson. He also wrote the first biography of his intimate friend, *Thoreau, the Poet-Naturalist* (1873, and enlarged, 1902).

CHANNING, William Ellery (1780-1842), the elder, graduated from Har-

vard at eighteen, and in 1803 became
pastor of the Federal Street Congregational Church in Boston, where he remained for the rest of his life. It is
difficult to reconcile the shy, retiring
nature of this semi-invalid with his historical position as the apostle of Transcendentalism (q.v.), the theologian who
perhaps more than anyone else dealt a
death blow to American Calvinism. His
sermon (1819) preached at the ordina-
tion of Jared Sparks marked the founding of the Unitarian movement. Both
Wordsworth and Coleridge were impressed when he visited them in England, and his influence on New England
writers exceeded that of any of his contemporaries. Not only did he infuse
American literature with a religious
liberalism, but he preached against the
"slavish imitation" of English models.
His collected *Works* were published in
6 volumes (1841-3) of essays and sermons noted for their grace, simplicity,
and reserved beauty.

CHAPMAN, John Jay (1862-1933), poet,
critic, translator, dramatist, essayist, lawyer, and amateur politician, graduated
from Harvard, was admitted to the New
York bar in 1888, and ten years later
left law to devote himself to literature.
One of the most brilliant figures in
America at the turn of the century, he
deserves a greater place in our literary history than has been accorded him,
but Chapman bore the curse of the man
too honest and perceptive to fit comfortably in either the literary or political
grooves of his time, while lacking the
force or conditions favorable to the cut-

ting of new grooves. In complaining
that traditional scholarship takes literature to be only an object of knowledge,
he was a precursor of the contemporary
"New Critics," but as Lionel Trilling observes in "The Sense of the Past," he
would also have decried their over-
emphasis on hard intellectual work and
concentration. His works include: *Emerson, and Other Essays* (1898), *Learning,
and Other Essays* (1910), *Greek Genius,
and Other Essays* (1915), *New Horizons
in American Life* (1932), and *Songs
and Poems* (1919).

CHASE, Mary Coyle (1907-   ), graduated from the University of Colorado,
and for many years thereafter, living in
Denver, she studied and tried the craft
of play writing until, in 1944, the production of *Harvey* (Pulitzer Prize, 1945)
made her famous and wealthy. Harvey,
a six-foot imaginary rabbit, is the constant companion of a genial inebriate
whose dream world is more attractive
and honest than the world of conventional reality. Mrs. Chase delineated a
child's dream world in *Mrs. McThing*
(1951), the title being the name of an
unseen witch behind the scenes that are
inhabited by fairy-tale gangsters and
respectable gossips. *Bernardine* (1952) is
about teen-agers. Mrs. Chase's work is
characterized by human comedy, intensified by a delicate fantasy applied to
the most unlikely situations and people.

CHAUNCEY, Charles (1705-87), opposed the *Great Awakening*, sponsored
by Jonathan Edwards, as a manifestation
of abnormal psychology in his *Season-*

able Thoughts on the State of Religion in New England (1743).

CHESNUTT, Charles Waddell (1858-1932), is generally credited as being the first Negro writer to use the short story as a serious medium of literary expression. *The Conjure Woman* (1899) contains dialect stories of the South told as though a white man were recounting the yarns of an old Negro friend, while *The Wife of His Youth and Other Stories of the Color Line* (1899) strikes out at the problems of his people both North and South. He is also the author of a biography, *Frederick Douglass* (1899), and three novels: *The House behind the Cedars* (1900), about a Negro girl's "passing" into white society; *The Marrow of Tradition* (1901), on race riots in a Southern city; and *The Colonel's Dream* (1905), concerning the hopelessness of racial hatred in a Southern town. Born in Cleveland, Chesnutt returned to the home state of his parents in 1874 and for more than a decade taught in the North Carolina schools. Later he established himself in the legal and business world of Cleveland. As a Negro writer, he was a pioneer in depicting the struggles of his race, during the postwar years, in works regarded not only as important social treatises but also as artistic accomplishments.

CHILD, Francis James (1825-96), was a major American philologist, an early student of Chaucer's metrics, and author of the valuable *English and Scottish Popular Ballads* (5 vols. 1883-98).

CHIVERS, Thomas Holley (1807-58), son of a prosperous Georgian planter, was named by Poe as "one of the best and one of the worst poets in America." He charged that Poe's "The Raven" was plagiarized from his own "The Lost Pleiad." His uncritical attempts to be different, as in comparing his broken heart to a broken egg, coining new words, and experimenting with metrics, might have established him as a precursor of the moderns except for his sentimentality and complete lack of discrimination. But he did influence Swinburne, Rossetti, and Poe—who influenced the symbolists, so that he still has some small claim to the distinction of influence.

CHOPIN, Kate [O'Flaherty] (1851-1904), was graduated from the Sacred Heart Convent in St. Louis, and in 1870 married Oscar Chopin, a distant cousin, who took her to New Orleans after a European honeymoon. Later they moved to a cotton plantation where Mrs. Chopin came to know the "Cajun" and Creole culture that provided the material of her novels and stories. In 1882 Oscar Chopin died of swamp fever, and after trying for two years to run both the plantation and her six children, Mrs. Chopin returned to her St. Louis home. Encouraged to write by friends who had long admired her letters, she made an intensive study of contemporary French writers, which perhaps partly accounted for her polish, restraint, and objectivity. Her stories remind the reader particularly of Maupassant. *At Fault* (1890) is an apprentice novel of Creole life, but with *Bayou Folk* (1894) and *A Night in Arcadie* (1897) she achieved her poignant style, her delicate humour. With the publication of her great novel, *The Awakening* (1899) a storm of invective arose against its presentation of mixed marriage and adultery, a cry that stifled the creative impulses of the sensitive author, a devout Catholic. This last novel, except for its rather primitive simplicity and economy of material, might be compared to *Madame Bovary*, so sensitive is the author's perception of the feminine mind divorced from its conventional face, and so capably does she arrange her unobtrusive architecture. Almost forgotten today, her few remaining books selling for fantastic sums, Kate Chopin deserves to rank as one of the distinguished novelists of the 19th century, and as a pioneer in the creative presentation of psychological realism in America.

CHURCHILL, Winston (1871 - 1947), was a prominent athlete at Annapolis, from which he graduated with high honors. Resigning his commission, he engaged in journalism for a year before turning his full time to the study of American history and ideals which he presented in a series of historical novels, including *Richard Carvel* (1899), *The Crisis* (1901) and *The Crossing* (1904), perhaps his finest work. Later, residing in New Hampshire where he was a defeated candidate for governor, his interest in contemporary politics was reflected in his novels about social and economic conditions.

CIARDI, John (1916-   ), grew up in Medford, Massachusetts. He took his B. A. degree at Tufts College and his M. A. at the University of Michigan, where his *Homeward to America* (1940) was awarded one of the largest Avery Hopwood awards ever given for poetry. During the Second World War he was a Central Fire Control gunner on a B-29, an experience interpreted in *Other Skies* (1947) with frequent philosophical comparisons of the military to civilian experience. *Live Another Day* (1949) reports his return to civilian life. He has also made one of the better translations of Dante's *Inferno* (1954). His poems are direct and clear interpretations of his personal experiences.

CLARK, Walter Van Tilburg (1909-   ), wrote *The Ox-Bow Incident* (1940) a story about an early lynching in his native Nevada, and *The City of Trembling Leaves* (1945) concerning a boy who aspires to become a composer. *The Track of the Cat* (1949) is a symbolic story of three brothers tracking a marauding mountain lion. *The Watchful Gods* (1950) is a collection of his shorter work, including his prize winning stories.

CLEMENS, Samuel Langhorne; See "TWAIN, MARK."

COATES, Robert M[yron] (1897 -), is perhaps best known for his morbid stories published in *The New Yorker*. *All the Year Round* (1943) is a collection of his stories. His novels include *The Outlaw Years* (1930), *Yesterday's Burdens* (1933) and *The Bitter Season* (1946), the latter portraying the fear and guilt of New Yorkers before the invasions that set off World War II. *Wisteria Cottage* (1948) is concerned with American neurosis at mid-century. Mr. Coates has served on the staffs of the New York *Times* and *The New Yorker*.

COBB, Irvin S[hrewsbury] (1876-1944), Kentucky humorist, wrote local-color stories featuring kindly, old Judge Priest ever befriending the needy. After magazine publication they were collected in *Old Judge Priest* (1915).

COBB, Sylvanus, Jr. (1823-87), born in Maine, the son of a Universalist minister, is generally considered to have been the first American to engage professionally in the mass production of fiction. He wrote more than 300 novellas and more than 1,000 stories of moral, sentimental, and sensational nature.

COBBETT, William (1763 - 1835), ran away from his home in Surrey at 14 and joined the British army. When his unproved exposés of army frauds resulted in law suits, he fled to the U.S. (1792), opened a bookstore in Philadelphia, and became a violently partisan pamphleteer for the Federalist party. His charge that Dr. Rush "killed" Washington through inept medical practice forced him to flee back to England, where he changed from Tory to radical. He later returned to the United States to agitate for labor and agricultural reforms. His best known book, the *Life and Adventures of Peter Porcupine* (1796), describes his first sojourn in America.

COFFIN, Robert P[eter] Tristram (1892-), grew up in Maine, graduated from Bowdoin, did graduate work at Princeton, was a Rhodes Scholar at Oxford, and served with the AEF as a lieutenant. Since 1934 he has been Pierce Professor of English at Bowdoin. His sixth volume of poems, *Strange Holiness*, won the Pulitzer Prize in 1936, and several new volumes have appeared since the publication of his *Collected Poems* (1939). He has also written novels, including *Lost Paradise* (1934), an autobiographical account of his Maine boyhood, and *John Dawn* (1936). His poems usually present Maine life in a hearty, masculine, and clear style. He has attempted to recapture the oral qualities of poetry by engaging in extensive reading of his work combined with informal "talks," and feels that this experience is fully as valuable for him as for his enthusiastic audiences.

Communism in American literature is almost impossible to define because of the subtleties involved in definition, whether of ideology, tendency, or party affiliation. Certainly the expression of the ideology (in *Masses*, 1911-24, and *The New Masses*, 1926-), whether Marxian or Russian, has augmented the interest in proletarian literature (q.v.). However, no work produced under this stimulus and meriting critical recognition has revealed a total subservience to communist doctrine, or can claim greater propagandistic value for the communist cause than such criticisms of their contemporary society as Stephen Crane's *Maggie* (1893) or Twain's *The Myster-*

ious Stranger (1916). Since the very nature and demands of literature reject the dictatorship of doctrine, the rejection of strictly communist verse or fiction may be virtually automatic. But such rejection has been predicated, so far in our history, on literary rather than political criteria. (See also, CRITICISM, *Marxism.*)

COMSTOCK, Anthony (1844-1915), has become the symbol for bigotry in the suppression of literature judged obscene by conventional or at times reactionary standards. As secretary for the New York Society for the Suppression of Vice, which he founded, he destroyed 160 tons of books and pictures.

Connecticut Wits (late 18th century), or Hartford Wits, was an informal association of Yale students and rectors, devoted to the modernization of the Yale curriculum and the declaration of the independence of American letters. However, they were essentially conservative in their attitudes, favoring orthodox Calvinism and the Federalist party. They collaborated in the production of *The Anarchiad* (1786-7), *The Echo* (1791-1805) and *The Political Greenhouse* (1799). Members included Theodore and Timothy Dwight, John Trumbull, and Joel Barlow.

CONNELLY, Marc[us Cook] (1890-), a Pennsylvania journalist, became a playwright after his successful collaboration with George S. Kaufman in producing *Dulcy* (1921), followed by *To The Ladies* (1922), *Merton of the Movies* (1922), and *Beggar on Horseback* (1924). His dramatization of Roark

Bradford's *Green Pastures* (1930) won the Pulitzer Prize, and he won the 1930 O. Henry Memorial Prize for "Coroner's Inquest."

COOKE, Rose Terry (1827-92), daughter of a wealthy Connecticut Congressman, developed a regional realism in writing short stories of the New England countryside. Her style was simple, humorous, and spontaneous, and although she achieved no deep penetration of character or dramatic situation, her stories give a realistic and vivid account of the period. Appearing first in *Harper's* and the *Atlantic Monthly,* they were later published as *Somebody's Neighbors* (1881), *Root-Bound* (1885), *The Sphinx's Children* (1886), and *Huckleberries Gathered from the New England Hills* (1891).

COOPER, James Fenimore (1789-1851), grew up in the wilderness of Otsego Lake, New York, where his father had taken up large holdings of land at Cooperstown. He attended school in Albany, and in 1802 entered Yale, but he was dismissed in his junior year for exploding a charge of gunpowder in the keyhole of a hallmaster's door. Following two years again in Cooperstown, he became a seaman, and a year later received his commission in the Navy, in which he served until 1811. In January of that year he had married Susan De Lancey, a potential heiress, and when his father was killed by a political opponent, Cooper retired to become a gentleman farmer, first at his father-in-law's estate at Mamaroneck, then for three years near Cooperstown. Here he was an aide-de-camp with the rank of colonel, an amateur architect, part owner of a whaling ship, and a secretary of the Agricultural Society, in which capacity he introduced Merino sheep to the area. In 1817 the growing family moved to a farm near Scarsdale which his wife had inherited, conveniently close to her family. He was 30 when, according to the now famous story of his daughter Susan, he threw down an English novel he had been reading aloud and declared that even he could write a better book. Challenged by his wife to prove it, he produced *Precaution* (1820), intentionally imitative of the genteel English novel of maners. Having served this cautious apprenticeship, he next wrote a novel closer to his own soil and awareness, *The Spy* (1821), which was an immediately popular success, although many critics complained of its crudities in style. With *The Pioneers* (1823), which sold 3500 copies the morning of publication, he began his "Leather-Stocking Tales," and introduced the famous Natty Bumppo in the scout's sixtieth year. He next utilized his knowledge of seamanship and the navy in *The Pilot* (1823). Now recognized as a leading American author, he moved to New York City, where he founded and presided over The Bread and Cheese Club. In *The Last of the Mohicans* (1826) he had an opportunity to present Natty Bumppo in his prime of manhood, and in this novel he first really exploited the Indians as a popular subject for fiction. Now, with six novels behind him, he took up a six-year

residence abroad, where he met Sir Walter Scott, his European rival in the romance of adventure. In Paris he became intimate with the aging Lafayette. When he returned to Cooperstown in 1833, he was repelled by what seemed to him the abuses of democracy; he quarrelled with neighbors over property rights, and brought libel suits against reviewers of his novels. His romantic style now became heavy with satire. Family financial difficulties, including debts left by his elder brothers added to his own extravagances, reduced him to writing for a living. *The American Democrat* (1838) expressed his growing aristocratic conservatism (already noticeable in the Tory sympathies that had crept into *The Spy*), and the book contains a brilliant defense of the rights of minority taste. *The Pathfinder* (1840) and *The Deerslayer* (1841) completed the Leather-Stocking series. Until 1850 he continued to literally pour out novels, histories and factual reports. The Littlepage Manuscripts, *Satanstoe* (1845), *The Chainbearer* (1845), and *The Redskins* (1846), chronicle the manners of New York and have as their subject the Anti-Rent War, a conflict between the propertied classes and the renters, in which Cooper again stood with the conservatives. His works have been the subject of considerable debate, some critics disposing of him as a romancer for boys, others taking the opposite extreme in seeing him as an American Homer. Several experts, including General Lewis Cass, have criticized his naiveté concerning Indians, and Mark Twain used both the Indians and the many impossible situations as the butt of a hilarious attack. Yet his work has great vitality and considerable authenticity in depicting the spirit of his various scenes. When more attention is paid to his later portraits of society, such as *Satanstoe*, and less to the adventure stories commonly associated with his name, he gains in critical esteem.

CORWIN, Norman [Lewis] (1910-   ), grew up in Boston, rejected his family's offer of a college education as wasted time for the serious writer, and established himself in New York as one of the rare champions of quality radio production. Writing and producing for the "Columbia Workshop," in the late thirties, he developed new techniques to combine the skills of engineering and poetry. His radio dramas have been published as *Thirteen by Corwin* (1942), *More by Corwin* (1944), and *On a Note of Triumph* (1945).

COTTON, John (1584-1652), a Cambridge dean and Puritan minister, moved to New England in 1633 as a result of a criticism of his nonconformity by the Archbishop of Canterbury. Here he became a church leader, and his *The Keyes of the Kingdom of Heaven* (1644) was the standard guide of the church until 1648. *The Way of the Congregational Churches Cleared* (1648) emphasized the aristocratic aspects of Calvinism. In 1647 he answered Roger Williams' "The Bloudy Tenet of Persecution" (1644) with "The Bloudy Tenent Washed and Made White in the Bloud

of the Lamb." His daughter married Increase Mather, and Cotton Mather was his grandson.

COWLEY, Malcolm (1898-   ), was an editor of the *Harvard Advocate,* served with an ambulance unit in France, studied for two years at the University of Montpellier, and held editorial positions in both New York and Paris. In 1930 he joined the staff of *The New Republic.* Although known principally as an editor and scholar, he has published two volumes of poems, *Blue Juanita* (1929), and *The Dry Season* (1942). A defender of contemporary American literature, he feels that good esthetics give a work sociological validity. In *Exile's Return* (1934) he attempts to analyze the post-war generation.

COZZENS, James Gould (1903-   ), grew up on Staten Island, N. Y., and published his first novel while a sophomore at Harvard, concerning the *Confusion* (1924) of an over-educated French girl. *Cock Pit* (1928) and *The Son of Perdition* (1929), although different in tone, are set in Cuba where he served as a teacher for the children of American engineers. *Castaway* (1934) is a fantasy about a man lost in a department store. His novel concerning a Florida air force base, *Guard of Honor*, was awarded the Pulitzer Prize in 1949.

CRANE, [Harold] Hart (1899 - 1932), planned his poem, *The Bridge* (1930), as an answer to T. S. Eliot's pejorative delineation of American culture. The son of a wealthy Cleveland manufacturer,

Crane never seemed to find himself completely at home in any milieu of the society he defended, although he sampled a variety. He worked in a munitions plant, at Brentano's bookstore, in a shipping yard, in a New York advertising agency, and on a newspaper and magazine. He sojourned on the Isles of Pines off Cuba, he toured England and France, and he sampled the literary association of both Pound and Tate. The climax of this condensed life was the award of a Guggenheim fellowship which took him to Mexico. Coming home by steamer, depressed and despairing of his genius, he dropped over the side into the Gulf of Mexico. His unhappy life and the desperation with which he wrote may have derived from disagreements with his family, which left him dependent on friends or on his own meager ability to sustain himself in the commercial world. In defending his work against charges of obscurity he declared that poetry should be appraised in terms of its own type of irrational experience and poetic logic of metaphor. His first collection of poems was *White Buildings* (1926). The *Collected Poems* (1933) were published posthumously, and contain many works that were not printed during his lifetime.

CRANE, Stephen (1871-1900), was the fourteenth child of a Methodist minister whose many assignments took the family to various New York and New Jersey communities before he died in 1880. Crane financed a year of study at Lafayette College by serving as correspondent for the New York *Tribune,* and another year at Syracuse University where he distinguished himself only as captain of the baseball team. When he was eighteen, perhaps because of his mother's death, he left college, and for two years reported, camped, played baseball, or loafed. At twenty he settled in New York and for three years he starved in the Bowery, reporting intermittently for the *Tribune* and *Herald.* His penetrating observations of life on Manhattan's lower East Side formed the basis for *Maggie: A Girl of the Streets* (1893), the first distinctively modern realistic American novel. Maggie's widowed mother drowns the sorrows of poverty in whatever drink she can come by, while advising her son and daughter to seek consolation in the Church. Hungry for the love which her home has never provided, Maggie succumbs to the advances of a young blade, but eventually does find consolation—in the East River. Although similar events had been presented in many an early American sentimental novel, in which the suicide erased any moral stigma, Crane's objective presentation left society no defense, and prospective publishers found it too grim. Actually, the book derives considerable humor from the mother's ethical solecisms, but this was thirty years before Sean O'Casey's similar *Juno and the Paycock* caused riots in a Dublin theater. Crane borrowed $700 from a brother to publish *Maggie* himself. Only a hundred copies were sold, and he burned the remainder for fuel during the following winter. Hamlin Garland showed a copy to William Dean Howells, and

Crane realized a small stir of interest from his investment. In December, 1894, he managed to sell *The Red Badge of Courage* to a Philadelphia newspaper for $90. This realistic, psychological study of a young soldier's fear and courage in Civil War battles caught the public fancy, and Crane found himself famous. The fact that an author who had never been in uniform could portray the warrior's emotions so accurately has caused much tedious wonder, but Crane had studied the histories well; he could use football as an emotional analogy, and in Tolstoy he found an artist's reaction to battle. *Maggie* was reissued, and his poems, influenced by the Bible as much as by Emily Dickinson, were published as *The Black Riders* (1895), followed by a collection of Civil War stories, *The Little Regiment* (1896). A syndicate sent him to Mexico as travel-correspondent, and upon his return in 1896 he was assigned a trip to Cuba. Shipwrecked enroute, he spent four days in a life boat, which he reported in the title story of a later collection, *The Open Boat* (1898). An assignment to the Graeco-Turkish front resulted in a satiric novel about a war correspondent, *Active Service* (1899). The suffering at sea, added to his earlier privations, left Crane in poor health. His exhaustion led to reports that he was a dope addict, while the realism of *Maggie* suggested to New York gossip mongers that he was a seducer. Completely disgusted, Crane moved to England, and there married his loyal wife, a former madam, who had nursed him in Florida after the shipwreck. Living in Surrey, he became a

close friend of Conrad. Covering the Spanish American War was to be his last assignment, for he became increasingly ill, and while resting in the Black Forest he died of tuberculosis at the age of twenty-eight and a half. His *Collected Works* were published in 12 volumes (1925-6).

CRAPSEY, Adelaide (1878-1914), wrote a small book of *Verse* (1915) patterned on the Japanese *hokku*, in the five-line stanza which she called a *cinquain*.

CRAWFORD, Francis Marion (1854-1909), was born in Italy, the son of Thomas Crawford, sculptor of the "Freedom" piece on the dome of the national capitol. Francis was educated both in America and Europe, and developed into a Henry James type of cosmopolitan, living most of his adult life at the magnificent Villa Crawford in Sorento, speaking numerous languages, and impressing the ladies with his tall, handsome physique. His romantic novels of India or Italy, designed purely for entertainment, were both popular and remunerative. The best of his forty-odd novels are *Saracinesca* (1887), *Sant' Ilario* (1889), and *Don Orsino* (1892).

CRAYON, Geoffrey, pseudonym used by Irving (q. v.) in publishing the three volumes of *The Crayon Miscellany* (1835).

CRÈVECŒUR, Michel-Guillaume Jean de (1735-1813), better known as St. John de Crevecoeur, was born in France, and came to America by way of England and Canada to settle on a farm in Orange County, New York, where he conducted agricultural experiments. It was here that he wrote his delightful *Letters From an American Farmer* (1782), so refreshingly descriptive of colonial life. Loyalist sympathies forced his return to England and then to France in 1780. Later he served as French consul in New York.

CRITICISM IN AMERICA:

Of all the literary arts, criticism quite naturally appears last in any national culture, for the critic needs a body of work as his subject, and even when he deals with theory, he will feel the need for some frame of literary reference within the culture he treats. There were reviews and commentaries in many of the early American magazines, but when the reviewer merely tells whether he likes a book, or whether he believes that young ladies should be permitted to read it, without any analysis, he can not be called a critic. Scholars are therefore inclined to list Edgar Allan Poe as the first American critic. He was the first to evolve a critical system based on his own analyses of the relationship between literature and life. Since Poe stood alone in this enterprise before 1850, and was followed in the next fifty years by other individualists who could not agree, together, on a system, many observers take the position that America did not really have a literary criticism of its own before the 20th century. But there were critical attitudes, even during the 18th century, and sometimes they were expressed in essays.

Eighteenth Century: The earliest attitude reflected the neo-Classicism of Dryden, Addison, and Pope, modified by Puritan didacticism. With the approach of the Revolutionary War nationalism began to challenge British neo-Classicism to a slight degree, and John Trumbull advocated a break with the "luxurious effeminacy" of European writers in his *Essay on the Use and Advantages of the Fine Arts* (1770). The neo-Classical emphasis on elegance and propriety was continued, however, in the lectures of Timothy Dwight, John Witherspoon, and John Quincy Adams, published during the first decade of the 19th century. Bryant's attack on the imitators of the sickly neo-Classicists (*Early American Verse*, 1818) was hardly a nationalist manifesto, since Wordsworth had originated the attack in 1800, and he and Coleridge were familiar to American readers. Coleridge's interpretations of the new German criticism were presented in the *North American Review* (1815-1939), and they helped to keep American attitudes parallel to the Transcendentalism which was developing in England. Discussions of nationalism during this period were directed more to considerations of language than literature. The American speech, vocabulary, and spelling already were noticeably different from those of the English, who became increasingly apprehensive of the effect that crude "Americanisms" might have on their own pure language. Much of the American speech, far from being an innovation as the British charged, was pure 16-century English, which had been retained from the early colonizers while London English changed.

Nineteenth Century: Transcendentalism as a literary attitude was much closer to American political mores than neo-Classicism had been. For the emphasis on rules and tradition it substituted a concern for the soul of the individual, and it was an essentially liberal philosophy. In America it developed more as a religious than a literary philosophy, however, and provided the basis for the Unitarian faith. Transcendental criticism was expounded by W. E. Channing, R. H. Dana, Sr., Henry Reed, Theodore Parker, and Bronson Alcott, but the major champion of the movement was to be Ralph Waldo Emerson.

Meanwhile, Edgar Allan Poe had challenged the early Transcendentalists and the Boston literary attitudes in general. His was an isolated voice, for he fit into none of the patterns that were developing and he attracted no disciples. Some of his ideas were neo-Classical. Like Sir Joshua Reynolds he ridiculed the notion that poetry derives from inspiration, and in *The Philosophy of Composition* he described his deliberate and mechanical decisions which shaped "The Raven." In *The Poetic Principle,* however, he attacked the Bostonian assumption, retained from Colonial neo-Classicism, that morality must be the purpose of a poem. He declared that "Just as the Intellect concerns itself with Truth, so Taste informs us of the Beautiful, while the Moral Sense is regardful of Duty. Of this latter, while Conscience teaches the obligation, and Reason the expediency, Taste contents herself with displaying the charms, waging war upon Vice solely on the ground of her deformity, her disproportion, her animosity to the fitting, to the appropriate, to the harmonious, in a word, to Beauty." Analytical criticism of this variety, and especially the distinction between beauty and morality, was new to America in 1846, and rather desperately needed. Literature is not necessarily amoral—the question of its moral content is still being argued; but the 19th century wanted to keep literature subservient to the moral platitudes, and this subservience often resulted in the production of immoral books. (See Novel in America).

The Transcendentalists seldom were analytical, and took no specific stand on the distinction between Duty and Beauty. Their mystic philosophy was a monism which held that everything in nature (i.e., the material world) is a microcosm containing within itself the meaning and the laws of existence. Channing asked for a poetry "which pierces beneath the exterior of life to the depths of the soul, and which lays open its mysterious working, borrowing from the whole outward creation fresh images and correspondences with which to illuminate the secrets of the world within us." Kant, the source of this philosophy, had defined transcendental knowledge as being concerned "not with objects, but with our mode of knowing objects so far as this is possible *a priori*" (*Critique of Practical Reason,* 1788). Thus there would seem to be neither argument nor agreement between Poe and the early Transcendentalists.

Later, Emerson did oppose Poe's poetry, and by implication, his theory.

Mentioning a "recent writer of Lyrics, his head a music box of delicate tunes and rhythms," as one who was not an "eternal" poet, Emerson declared that it is not meters but a meter-making argument that makes a poem. His literary notions permeate all of his essays, philosophical as well as critical, and his criticism usually is a literary branch of his transcendental idealism. He felt that critics should be poets dealing with the essential quality of mind of their subjects, so as to teach the reader to partake at first-hand of the spirit which motivated the genius. When he became more specific concerning his requirements of the poet, he asked for an American with a tyrannous eye, who would see in the barbarism and materialism of the times another carnival of the gods of Homer. He declared that America is a poem in our eyes; its ample geography dazzles the imagination, and it will not wait long for meters. Thus he urged a break with conventional prosody. Curiously, Emerson rarely practised what he preached, and it remained for Whitman to embarrass the older sage by exceeding the bounds of "taste" in his New York carnival of the gods of Homer.

Unlike Emerson, Poe, or Lowell, Whitman had a self-administered and disorganized education, and his criticism is less a system than an attitude. He went even further than Emerson in declaring that America is unrhymed poetry, and that the country would produce a bard commensurate with her great people and ideals. He advised the poet to re-examine everything he had been told at school, in church, or in books. The poet is the lover of the known universe. No elegance or attempts at originality should be allowed to mar his clarity. By following the precepts of the American ideal, despising riches and tyranny, loving fellow men regardless of station, the poet could become great.

James Russell Lowell, recognized as one of the greatest critics in American literature, was a professor of comparative literature at Harvard, and he had the scholar's background and scope. In his early criticism he advocated a didactic poetry to be enlisted for the cause of the abolition of slavery; in his second period (c. 1850-c. 1867) he advocated an indigenous national poetry; but in his mature period he advocated a literature, however national its roots, which would be devoted to the universal and unchanging aspects of the human spirit. He defined form as the artistic sense of decorum controlling the co-ordination of parts and insuring their harmonious subservience to a common end. For him, imagination was the common sense of the invisible world, as understanding is that of the visible. Lowell even contributed his bit to the tradition of Shakespeare criticism: Shakespeare's characters are the most natural ever created, but they are also representative of the ideal in man, and are more truly the personifications of abstract thoughts and passions than those of any allegorical writer. He felt that the age of Shakespeare was richer than his own only because it had such a pair of eyes to see it, and he agreed with Emerson and Whitman that the American scene, including railroads and steamships, would be a proper subject for the proper poet. But he exceeded them in mature wisdom when he predicted: "Until America has learned to love art, not as an amusement, not as the mere ornament of her cities, not as a superstition of what is *comme il faut* for a great nation, but for its humanizing and ennobling energy, for its power of making men better by arousing in them a perception of their own instincts for what is beautiful, and therefore sacred and religious, and an eternal rebuke of the base and worldly, she will not have succeeded in that high sense which alone makes a nation out of a people, and raises it from a dead name to a living power." Another indication of Lowell's maturity as a critic is revealed in the fact that, while he recognized the dangers of American materialism, he did not become hysterical in damning this trend. If America was commercial, so also, he pointed out, were renaissance Florence and Shakespeare's London. In his *Biglow Papers* he praised rustic common sense and criticized "bookish" writing, giving an impetus to the new movement of regionalism.

Regionalism, perhaps an inevitable consequence of the national development, was a recognition by editors as well as writers that interpretations of purely local characters and situations had a place in our literature. Hippolyte Taine, the French critic, had considered literature to be an index of a nation's social history, since it was determined by the author's time, place, and race. It is difficult to determine what effect Taine may have had on American region-alism, but he was read by many Americans, and probably gave the movement an additional impetus.

Realism: One of the critics under the influence of Taine was William Dean Howells, the high priest of American Realism in its early stages. Howells was influenced by Tolstoy's ethics, as well as by Taine's determinism. This mixture was blended with his own happy experience of rising from obscurity to fame at a fairly easy pace. Standing in complete contradiction to the mild Classicism of Lowell, he stated that the Realist must objectively report reality, neither idealizing nor selecting. He ridiculed the disciples of both Plato and Aristotle for imitating an imaginative synthesis of reality, stating that this was analogous to reproducing a cardboard grasshopper when a real grasshopper was available. The critic should examine a work of literature as the botanist examines a plant, or as the zoologist examines a grasshopper. A zoologist can not evaluate a grasshopper, he can only describe its function, class and character, and how it is imperfect or irregular as a representative of its species. This is Taine's determinism. It would appear to leave no opening for aesthetics in the creation of literature, but Howells solved this problem by borrowing Tolstoy's conception of ethical growth as a part of the scientific evolution. The perfect aesthetics in a novel results from the perfect ethics, the artist and moralist working together for righteousness. Howells lacked a scholarly training, being, like Whitman, mainly self-educated, and there are many inconsistencies in his critical philosophy which reveal the absence of a disciplined organization of ideas. In his own work, scientific objectivity suffered much more from his emphasis on ethics than it suffered from organization and plan in the work of his contemporary, Henry James. In decrying the poets' imitations of each other rather than nature, he declared that he could judge but poorly of anything while he measured it by no other standard than itself, a statement that is in direct contradiction to his botany or grasshopper analogies. And while his theory abrogated judgment, no critic judged more than did Howells.

Henry James was a Realist who operated in the more classical tradition of Lowell and emphasized aesthetics as a means of achieving morality rather than the reverse. For him the air of reality, solidity, and specification was the supreme virtue of the novel; yet experience was "an immense sensibility." Whereas Howells attacked English criticism for instructing a man to think that what he likes is good, instead of instructing him first to distinguish what is good before he likes it, James denied that there are certain things which people ought to like and can be made to like. For James, the ultimate, the ancient test, is whether the man likes a work. He covered morality to his own satisfaction by declaring that "no good novel will ever come from a superficial mind." Questions of art, for James, were questions of execution, not of morality, subject matter, optimism or pessimism, or type. He considered art to be mainly selection, but the selection needed to be

inclusive as well as typical, and artificial rearrangements were to be avoided. An independent and objective thinker, his criticism as well as his art of the novel was made up of all of the ideas available to him, stemming from Taine, Sainte-Beuve, Zola, and his brother William's Pragmatism; yet he was subservient to none of these sources. He recognized that the impression of reality could best be secured through its artistic manipulation.

In the last decades of the 19th century the critics paid their closest attention to the novel, primarily because most of the active critics were novelists, but also because the new interest in Realism could best be exploited through consideration of the novel form. Although Whitman is often considered to have been a pioneer in realistic poetry, and later poetry was inevitably conditioned by the new Realism, the more artificial nature of poetry reduces its usefulness as a frame of reference concerning representations of true life.

The trend toward Realism in the novel was stimulated by the new interest and developments in science. Zola had used Claude Bernard's extension of the experimental method to medicine as a basis for his demand that literature also be empirical. In 1894 Hamlin Garland published his *Crumbling Idols* to show that Darwinism had inspired a revolt against tradition in literature as well as in life and religion. Truth became more important than beauty, and its importance derived from its potential as a weapon against social injustices. If literature could not be as empirical as Zola had demanded, it could at least be the hand-maiden of the social sciences.

Early Twentieth Century: A conflict quite naturally developed after 1900 between the traditionalists and those who were active in or sympathetic with experimental approaches to literature. It is a frustrating aspect of the study of criticism that the battle lines usually are drawn up after the battles are history. Thus the dispute, or, as it is some-times called, "the battle of the books," was not a clear dichotomy between the Realists and the Romanticists, as some text books lead us to assume. Realism itself is so diversified a classification that it is broken down into four different tempers: (1) the realistic temper, aim-ing at exact representationalism; (2) Naturalism, the scientific study of the evils of society revealed in narrative; (3) Impressionism, which intensifies re-ality by producing the impression of experience rather than its actuality; and (4) Expressionism, which departs radi-cally from reality using symbolism, psy-chology, and fanciful narrative tech-niques to apprehend the ultimate rather than the visual truth. Twentieth-century writers have shown a disturbing tend-ency, however, to play hop-skip-and-jump around these categories, sometimes even jumping back into Romanticism. Twentieth-century critics were so busy endeavouring to work out a system to handle the abundant new material that they seldom were in a position to ap-praise it.

The critical battle, initially, divided the traditionalists, the conservatives, from the experimental radicals. Among the conservatives stood G. E. Woodbury, W. C. Brownell, W. P. Trent, P. E. More, and Irving Babbitt. The radicals included Brander Matthews, T. R. Louns-bury, F. L. Pattee, James Huneker, Van Wyck Brooks, H. L. Mencken, Lewis Mumford, Waldo Frank, and Randolph Bourne. Both groups were seeking a philosophy not only of literary values, but of social values as well. Thus Bourne exhausted his energies, before an early death, in criticizing what he saw as the false rationalization of the American cause during the First World War. Thus Brooks based his critical ex-amination of literature on a critical ex-amination of our Puritan heritage, which he saw, in partnership with practical materialism, as the greatest enemy of the American artist. The conservatives, also, were disturbed by scientific materialism, and related the political situation to lit-erature. In *Democracy and Leadership* (1924), Irving Babbitt favored human-istic and aristocratic democracy over what he termed the false theory of nat-ural rights. H. L. Mencken, the most radical of the radicals in his examina-tions of literature, had one foot in each of the camps when he attacked society: he stood with Brooks in his quarrel with the Christian moral code; he joined Bab-bitt and the Humanists in deploring the lack of a "civilized aristocracy" in our society. Mencken had found his formula for treating aesthetic, ethical, and social questions in the philosophy of Nietzsche, particularly in *The Birth of Tragedy*, which balances Dionysus against Apollo in a perpetually conflicting duality. This was the duality that created Attic tragedy. For Mencken, examining American mores, it was the duality of dogma and doubt. It was a perfect duality, perfect beyond the intentions of Mencken, for classifying the two critical camps in the battle of the books.

There was less unity in the radical Dionysian camp than among the Apollonian conservatives. Bourne was an early leader of the radicals by reason of his aggressiveness, and upon his death, Brooks became the leader. But neither of these men attracted disciples who would carry forward their ideas with only slight modifications. The camp was united by its methodology, which was empirical criticism based on the social sciences, and also by a common antipathy to the restrictive mores of the past, both religious and academic. Some of the members, including Waldo Frank and John Reed—and Max Eastman in his earlier years, even moved off toward the dialectical materialism of Marx.

Humanism: The conservatives, who were also critical of the American tradition, believed that the cure could not come from the inductive reasoning of the social sciences, but must come from the inner conscience of the individual, from Humanism rather than from humanitarianism. William Crary Brownell believed that aesthetic judgments, to be complete, needed as their basis the Platonic doctrine of form, Aristotelian reason, and Christian ethics. The function of criticism, for Brownell, was to "discern and characterize the abstract qualities informing the concrete expression of the artist." Irving Babbitt carried Brownell's position forward, declaring

that the world would be better if people had made sure they were human before they attempted to be superhuman. Naturalism is in error, he declared, when it chooses the world of phenomena for its center. The Humanist aims at proportion through cultivation of the law of measure. In endeavoring to escape from a theological monism, mankind had fallen into a naturalistic monism. Babbitt presented a triangle of man, nature, and the divine, each in its proper proportion, as the ideal. Paul Elmer More was, in turn, a disciple of Babbitt, whom he had known as a fellow student at Harvard, and he criticized the Naturalists for being under the spell of *The Demon of the Absolute* (1928). Norman Foerster's *Humanism and America* (1930) helped to solidify Humanism as a movement at a time when it was declining in creative vitality. In summary, Humanism was in the tradition of English and American letters, and Matthew Arnold was its progenitor, as the Transcendentalists were its cousins. The rebels, however, usually took their ideas from Europe, especially from France.

Impressionism: In the midst of the battle of the books appeared a disciple of the Italian, Croce, who blandly declared that "we are done with" the very sort of criticism that was being produced by both camps, and would continue to be produced to some degree for the next four decades. Joel E. Spingarn was the American proponent of the critical system that is called Impressionism, which might be described as "the philosophical wisdom of aesthetics," as opposed to analytical systems of aesthetics.

It is represented by Goethe, by Carlyle in his essay on Goethe, by Sainte-Beuve, Lemaître, and Anatole France, and most philosophically by Croce. Spingarn agreed with France that the function of criticism is to have sensations in the presence of a work of art and to express them. He complained that the historical critic takes us away from the work in search of environment, age, race, and the poetic school of the artist; the psychological school sets us to work on the biography of the poet; the dogmatic critic sends us to the Greek dramatists and to Aristotle. But Spingarn wished to enjoy the poem itself. He saw as the eternal conflict of criticism the argument between his own group, the Impressionists, who emphasize enjoyment, and the dogmatists, who pass judgment. Advocating a kind of literary anarchy, Spingarn declared that we are done with all rules, with literary genres, with vague abstractions concerning the comic or the tragic or the sublime, with all moral judgment of literature, with technique as separate from art, with the *Kulturegeschichte* of Taine, with evolution in literature, and finally, with the old rupture between genius and taste. The creative and the critical instincts are one and the same. Spingarn's "The New Criticism" (1910) had little immediate effect on the schools of dogma. We were actually done with none of the dogmas he sought to bury. Vernon L. Parrington's *Main Currents in American Thought* (1927-1930) was an exhaustive study of literature in terms of political, social and economic history. Marxist critics continued to appear on the scene

with varying degrees of dogma, and the disciples of what later came to be called "the New Criticism" tended to be absolutist, whatever their special philosophy. Yet Spingarn may have helped divert criticism to some degree from the study of society to the study of aesthetics, may have prepared the way for such of the New Critics as T. S. Eliot, R. P. Blackmur, Allen Tate, John Crowe Ransom, and Yvor Winters. Spingarn also may have been responsible for the fact that Edmund Wilson and Kenneth Burke, who are classified as Marxist critics, were among the shrewdest investigators of the problems and examples of literary creation after 1930, and that they seldom became involved in the fantastic rationalizations of the British Marxist, Christopher Caudwell. It is, in fact, quite possible to read their best work without gaining any notion of their political sympathies. Even the less able Marxist critics who respected the "Stalinist" party line frequently insisted on some distinction between art and politics.

Marxism: Future historians of our literature may refuse to recognize the Marxism of the thirties as an integrated part of American criticism. With the stock market crash of October, 1929, and the following depression, national attention was focussed upon the faults of our economic system, or lack of system. Spingarn's "art for art's sake" was an embarrassingly precious attitude to hold in the face of this national emergency. Editors and writers alike were concerned with economic problems, whatever their economic ideology, and even as the new Humanists were marshalling their forces against sentimental humanitarianism, this very humanitarianism dominated the periodicals. It was inevitable that some critics as well as some journalists, writers, politicians, factory workers, union leaders, and even capitalists, should be attracted to the Marxist cure. The critics who chose this pattern included Granville Hicks, Michael Gold, V. F. Calverton, Joshua Kunitz, Newton Arvin, Philip Rahv, and James T. Farrell. Meanwhile, an earlier Marxist, Max Eastman, published *Artists in Uniform* (1934), revealing his disillusionment with orthodox Marxism, and both Farrell and Rahv warned against the regimentation of art. The movement ended abruptly when Stalin made a pact with Hitler in 1939.

The New Critics: While the early Humanists were fighting the social cynics and while the neo-Humanists were fighting uncritical humanitarianism, a new school of criticism was developing abroad, led by Ezra Pound and T. S. Eliot, and following in the tradition of those defenders of artistic autonomy who were so often neglected in their own day: Poe, Lowell, James, Santayana, Huneker, and Spingarn. Pound and Eliot were destined to move into their own distinct and curious social dogmas, but they and their disciples rescued literature from its subservience to social dogma. They operated within the American tradition when they developed a critical dualism of art and morality, but their morality was not the didacticism of the Transcendentalists. (The Marxists were the inheritors of Longfellow.) In many respects the new school

derived from Henry James, the greatest exponent of formal craftsmanship and aesthetic discipline (*The Art of the Novel*, 1907-09; collected, 1934), who at the same time believed that all art has a moral basis. Whether his rather special emphasis on morality or his general philosophy had been influenced by his father, a disciple of Swedenborg, is presently being argued by a few critics. (Quentin Anderson has just completed a study of this question.) Certainly the measure of his morality was not a product of New England, and had he written a *Comedy*, we could expect to find most of our Puritan forefathers in the eighth circle of his *Inferno*. He was concerned with the ultimate morality of ethics, taste, and integrity, and never with creed. It is doubtful that any of his belated disciples measured up to his measure of morality, but they were closer to it than the anti-Puritan Naturalists, the Marxists, or the Humanists.

James was an Anglo-American, and it was fitting that the earliest of the new critics, Pound and Eliot, should also be American expatriates. The New Criticism has been, from the beginning, an Anglo-American development, although it borrowed Paul Valéry's theme of contemporary spiritual decay. Bergson's distinction between faith and reason, and certain technical innovations of the French Symbolists. Like Humanism, it attempts to systematize the philosophy of Arnold, and Eliot is so close to being a Humanist that he was a contributor to Foerster's anthology, *Humanism and America* (1930). On the other hand, Eliot is Thomistic in matters of religion.

The eclecticism of the New Criticism makes it extremely difficult to classify in terms of any other movement. It derived its emphasis on form and morality from James, but its absolutism concerning taste runs contrary to James and parallel to Howells. It selects the English Metaphysical poets as an example of correctness in poetry, for their exact correlation of structure and idea, for their functional imagery, and for their irony and wit. The New Criticism repudiates the neo-Classicism of the Augustan period for its non-functional, decorative use of metaphor, and it repudiates 19th-century Romanticism for its sentimental idealism.

The movement may be dated from Pound's meeting with T. E. Hulme in London, in 1907. He also met James, W. B. Yeats, Ford Madox Ford, and Remy de Gourmont, and derived fresh insights from each of them, but Hulme's criticism of the Romantic spiritual disunity was to be the keystone in the New Criticism. Pound was not so much a systematic critic as a gadfly and organizer. He organized Imagism and Vorticism (qq.v.), and through his influence on Eliot, who came to England in 1913, he inadvertently launched the New Criticism. Thereafter Eliot developed the ideas of Hulme and the enthusiasms and disaffections of Pound into a systematic attack on Romanticism which became the basis for the New Criticism. Beginning with *The Sacred Wood* (1920), he produced a series of essays on society, religion, and art. (His *Selected Essays* (1932) combine his most important literary criticism.) His major

disciple in England was F. R. Leavis of Cambridge University, and the American parallel to Leavis has been Allen Tate. English critics who contributed to the New Criticism include Martin Turnell, William Empson, Herbert Read, John Middleton Murry, and I. A. Richards, although Murry and Richards can not, like the others, be defined as disciples. The Americans who have followed Eliot directly or through Tate are John Crowe Ransom, Robert Penn Warren, John Peale Bishop, R. P. Blackmur, and Cleanth Brooks. Yvor Winters also operates in this tradition, but selects quite different poets for his admiration, and places a greater emphasis on morality than does Tate.

Although the New Critics have minor differences of opinion and are known for their individual contributions to the system or exploitations of it, their agreement is so general that a single description of its tenets will serve them all. Whereas earlier 20th-century criticism concentrated on the novel, and was written mainly by novelists, the New Criticism constantly uses poetry as its example, and is written by poets. This writing is characterized by the use of special terms which are meaningless to the uninitiated. The most important term is, "dissociation of the modern sensibility." It is the premise of the New Critics that contemporary civilization is in a state of decay, and that this decay began with the rise of Protestantism and the weakening of the authority of the church. Thereafter, man's intellectual capabilities were devoted increasingly to science, and both religion and art were left only with emotion. The modern glorification of the scientific vision at the expense of the aesthetic vision, tradition, and religion accounts for our present dissociation of sensibility. This loss of the religious and social traditions deprives the poet of his moral and intellectual authority. The corresponding loss of a fixed convention deprives the poet of a unifying relation to his society, a rational structure of ideas which would provide him with discipline and authority in the organization of his creative work. Again, the loss of any belief in a homogeneous religion or mythology deprives the poet of his dramatic conception of human nature, for which he tends to substitute his own personality. However, the complexity and diversity of the contemporary world, which replace the old world order, leave the poet neither moral nor intellectual standards for measuring his own personality, leave him no "fixed point of reference" for his sensibility, and the result is intellectual chaos.

The New Critics believe that the Renaissance brought about a dichotomy between art, or qualitative knowledge, and science, defined as quantitative knowledge, followed by a reduction of all knowledge to the quantitative. The Renaissance also engendered a confusion of faith and reason, which Classicism saw as two distinct entities, and the confusion causes our spiritual disunity. In spiritual and ethical matters, the Renaissance substituted humanism for the absolute, faith. Classicism had postulated absolute values by which man was judged as limited and imperfect. The Renaissance humanists

ing beyond the poem to investigate, for developed Romanticism, which regards man as a reservoir of desirable possibilities. Not only does Romanticism cause spiritual disunity, but it dissipates the possibility of dramatizing the tragedy of the human soul, for it destroys the belief in absolute evil, an hypothesis which tragedy demands. Science, also, has devitalized religion and art, by monopolizing reason and reducing the spiritual and the aesthetic to irrelevant emotion. Poetry, Ransom maintains, is ontological; it presents, not simply emotion, but a total description of the object. The difference between science and poetry is not one of intellect and emotion, but of abstraction (science) and concrete specification (poetry). The great poem achieves what Tate calls a *tension* of *extension* (abstract idea) and *intension* (concrete image), a fusion of abstraction and concretion. Art apprehends and concentrates our experience within the limits of form, and the value of a poem is judged by the aesthetic structure of its single intention. The center of value, says Eliot, is not in the poet's feeling, but in the pattern he makes of his feeling. Similarly, enjoyment is not predicated upon agreement with the ideas expressed in the poem, but upon the relevance of the ideas to each other. Art proves nothing; it creates an absolute, coherent experience which it is impossible to achieve in living experience, since reality develops neither coherence nor form. This artistic experience results from the "signification" of a living experience, an emotion, or an idea. The "traditional (i.e. New) critic" finds noth-

there *is* nothing beyond the poem; therefore, the poem's origin, psychological effect upon the reader, or social value does not signify. Thus the New Critics reject "Platonic" (didactic) poetry, which confuses art with science, and advocate a "Physical" poetry which is concrete. The highest order of poetry is the Metaphysical, which synthesizes widely divergent and conflicting images of emotion and thought.

Since the relative values of emotion and thought have been argued for centuries, with thought construed as argument or knowledge or precept, and emotion construed as the non-analytical reaction of the reader, it is rather difficult for many students of criticism to conceive of an ontological examination of poetry, where both the intellectual and emotional aspects of the poem are examined in their correlation and not as separate entities. Seen in these terms, the reading of a poem is neither an intellectual nor an emotional experience, but simply an experience. It is comparable to a living experience, where intellect and emotion normally are intertwined; yet, as was stated earlier, it is more than a living experience, since life has neither coherence nor form. A "close reading" of the poem, rather than a mere reaction, is essential to the complete realization of that complete experience which the poem can provide. Obviously, the poem can not be judged until its potential of experience has been realized.

The reaction of scholars and other critics to the New Criticism tends to divide between unrestrained praise and uncritical hostility. Perhaps the move-

ment is still too new to receive an objective appraisal. Tentatively, it seems safe to judge that it has contributed a great service in helping to restore the autonomy of art. The emphasis on "close reading" of poems, both ancient and modern, as developed by Empson in England and Brooks and Warren in America, has led to a greater academic emphasis on the complete study of poetry. The philosophy of the New Critics has had a specific value for those aesthetes and artists who feel the need of system to identify personal experience with an orthodoxy of tradition and moral responsibility. The New Critics have provided a system for judgment and creation which has considerably improved our literature, and they have assisted the substitution of a modern convention of language for the dead conventions of the 19th century.

Critics of the New Criticism seek to qualify some of these recognized contributions. They point out that the new language was apparent in the work of Emily Dickinson, Robinson, Sandburg, Frost, and many another poet who wrote before the new influence was effective. Romanticism, they argue, was a fairly dead issue before the New Critics appeared on the literary scene. While recognizing the value of "close reading," some observers feel that the emphasis on a Metaphysical poetry has led to an over-emphasis on the poem's intellectual content which is as narrow as the romantic over-emphasis on emotion; consequently, the contemporary poet too often constructs "crossword puzzles" rather than passionate apprehensions of experience. The reading or deciphering of such poems becomes and remains strictly an intellectual exercise rather than an ontological experience.

The philosophy which the New Critics developed to rationalize their system, although not essential to use of the system, may perhaps become its most vulnerable part, and may be used unfortunately to discredit the whole. The New Critics' neo-Thomistic regimentation of history may itself seem sentimental in another hundred years if it becomes evident that scientific materialism has actually increased the artistic sensibility of modern mankind, taken as a whole. It may be asked, is quality alone, considered as an abstraction, the only criterion to be used in measuring man and society, or may we introduce a consideration of the quantity of quality? It is a fallacy of neo-Thomism to compare the vocal aristocracy of earlier periods with the vocal bourgeoisie of the present, forgetting that the latter would have been illiterate and grubbing peasants in the world of St. Thomas Aquinas. Henry James owed his leisure for aristocratic development to his grandfather, an Irish immigrant who amassed three million dollars in materialistic land speculation.

The optimism of the New Critics is restricted to their own accomplishment, but if their achievement in systematizing literary criticism has not been equalled in the history of literature, as they believe, and if it has permeated the contemporary study of literature, as is evident, America can hardly be suffering from intellectual chaos. In terms of our potential creativity, the New Criticism

can be dangerous by reason of its superb logic, as every dogma has been dangerous in the past, as the romantic dogma was dangerous to Dickinson and Robinson, and to Eliot when he first broke with what was accepted as tradition. The superb system may be forgotten, and the critics may reject anyone who does not write like them. It is to be noted that they did not welcome Dylan Thomas with enthusiasm.

Yvor Winters has been especially useful to modern criticism by reason of his heresies while operating within the new system. No contemporary critic can or should escape the influence of the movement which Eliot generated, but among such younger critics as Peter Viereck, Richard Chase, and Eric Bentley there is a refreshing individuality. Meanwhile, more traditional critics, Lionel Trilling in the field of the novel, and Joseph Wood Krutch in drama, continue to examine literature as if it were a matter of research and reason.

CROTHERS, Rachel (1878-), grew up in Bloomington, Illinois, and as a young woman went to New York to study the theater. In addition to acting, directing and producing, she tried her hand at writing, and became an immediate success with *The Three of Us* (1906). Thereafter she averaged one play a year until 1938, when *Susan and God* was produced. Most of her plays dealt with ethical problems of women, and her only financial failure was in producing a play for another woman playwright.

CROUSE, Russel (1893-   ), has collaborated mainly with Howard Lindsay in writing *Anything Goes* (1934), *Life With Father* (1939), and *State of the Union,* Pulitzer Prize winner in 1946.

CUMMINGS, E[dward] E[stlin] (1894-), took his B. A. and M. A. degrees at Harvard, served with an ambulance corps in France, and after the first World War settled in Paris to write and paint. He uses jazz rhythm, slang, and unorthodox typography to convey a cynical thesis concerning contemporary experience. His poetry has been published as *Tulips and Chimneys* (1923), *Vi Va* (1931), *No Thanks* (1935), and *1 x 1* (1944). Much of his work defies classification, and one of his books has no title (1930).

D

DALY, [John] Augustin (1838-99), wrote or adapted from the original French and German nearly a hundred plays, and produced still others in his New York theater.

DALY, Thomas Augustine (1871-1948), wrote light verse, the humor of which derived mainly from the dialect of his Irish-American and Italian-American characters: *Canzoni* (1906), *Carmina* (1909), *Madrigali* (1912), *McAroni Ballads* (1919) and *McAroni Medleys* (1931).

DANA, Charles Anderson (1819-97), New York journalist who edited the *American Cyclopedia* (1858-64) with George Ripley, and *The Household Book of Poetry* (1857). In his youth he taught Greek and German at Brook Farm (q. v.), and in his later years he published the New York *Sun,* crusading violently for his own opinions or prejudices.

DANA, Richard Henry, Sr. (1787-1879), was a founder and editor of the *North American Review,* in which he published Bryant's "Thanatopsis." As a romantic critic he opposed the rococo expressions so common then in poetry, and insisted that Shakespeare was a greater poet than Pope—a premise then considered debatable. Of the many poems he contributed to contemporary journals, *The Buccaneer* is best remembered. In later life, overshadowed by his son, he devoted himself to religious studies.

DANA, Richard Henry, Jr. (1815-82), son of R. H. Dana, Sr. (q. v.), left Harvard in his junior year because of eye trouble suffered in an attack of measles, and shipped as a common seaman to California by way of Cape Horn (1834). From the diaries of this trip he later wrote his vigorous and fairly realistic *Two Years Before the Mast* (1840), on which his reputation is based. Later Dana graduated from Harvard law school, and devoted his life principally to politics, in which he never won the positions to which he aspired. His manual, *The Seaman's Friend* (1841), incorporates legal advice for sailors, and reflects the sympathy for their cause he had shown in his novel.

DANIELS, Jonathan Worth (1902-   ), is equally at home in politics, journalism, and literature. He attended the University of North Carolina. Since 1933 he has been editor of the Raleigh (N. C.) *News and Observer* except during the war years when he was in Washington as an adviser to Presidents Roosevelt and Truman. *Clash of Angels* (1930) is a novel based on mythology. His books of social history are noted for their perception and stylistic excellence: *A Southerner Discovers the South* (1938); *A Southerner Discovers New England* (1940); *Tar Heels* (1941); *Frontier on the Potomac* (1946); and *The End of Innocence* (1954). *The Man of Independence* (1950) is a biography of Harry S. Truman. The father of Jonathan Daniels was Josephus Daniels (1862-1948), Secretary of the Navy under President Woodrow Wilson and later Ambassador to Mexico. He was the author of nine books, notably the five-volume autobiography beginning with *Tar Heel Editor* (1939).

DARGAN, Olive Tilford, a writer who is particularly known for her fictional deromanticizing of the Southern Highlanders, has treated her subjects sympathetically and poetically in varied mediums. Born in Kentucky, she first taught school, later attended Radcliffe, then moved to the North Carolina mountains with her husband. There she has lived since his death in 1916. After three volumes of poetic closet dramas, she published three books of poetry, *Path Flower* (1914), *The Cycle's Rim* (1916), and *Lute and Furrow* (1922). A book of short stories *Highland Annals* (1925) was revised as *From My Highest Hill* (1941). Under the pseudonym of Fielding Burke, she wrote several novels: *Call Home the Heart* (1932), *A Stone Came Rolling* (1935), and *Sons of the*

Stranger (1947). The first two treat the mountaineers caught in the violent industrial struggles in North Carolina during the 1920's.

DAVIS, Clyde Brion (1895-   ), a western journalist who turned to professional writing with the success of *The Anointed* (1937), the story of a sailor navigating his own mind. *The Great American Novel* (1938) concerns a newspaperman who hopes to write the great American novel. His other novels include *Nebraska Coast* (1939), *Sullivan* (1940), *The Stars Incline* (1946) and *Playtime Is Over* (1949).

DAVIS, H[arold] L[enoir] (1896-   ), was born in Oregon, and became a printer's devil at nine, a sheepherder at ten, a cattle puncher at eleven, a schoolboy at twelve, and a deputy county sheriff at seventeen. As a boy he learned Spanish from Mexican sheepshearers, and after graduating from high school he continued his self-education, eventually becoming a linguist, an authority on western culture, and a distinguished student of Spanish literature. Several of his poems were published in *Poetry* before 1920, and were later collected as *Proud Riders* (1942). He won both the Harper prize and the Pulitzer Prize for his first novel, *Honey in the Horn* (1935), an unusual love story concerning a teen-aged boy and girl of heroic dimensions migrating back and forth across Oregon, told with a rare, human humor, and rich in its background of lusty characters searching for the promised land in Oregon between 1906 and 1908. *Harp* *of a Thousand Strings* (1947) places three American sailors in Tripoli during the Barbary War to contrast the European with the American way of life. *Beulah Land* (1949) again presents two amazingly self-reliant youngsters in a gradual migration from the Cherokee country of North Carolina to Indian territory in Oklahoma. *Winds of Morning* (1952) employs an old horse herder and a young assistant sheriff to contrast two different epochs in the history of Oregon. His short stories were collected in *Team Bells Woke Me* (1953). Davis now lives at Point Richmond on San Francisco Bay in a grove of eucalyptus trees planted by Jack London, often spending time in Mexico, where he lived for five years, partly on a Guggenheim Fellowship, while writing *Honey in the Horn*. At present he is working in Hollywood to prepare an ancient movie, *The Covered Wagon*, for a reproduction compatible with its exaggerated reputation. Davis has a natural talent for story telling and this facility, added to the prejudice against taking western stories seriously, often blinds the reader to his many levels of symbolic overtone.

DAVIS, Owen (1874-), Harvard graduate of 1893, won the Pulitzer Prize for his play, *Icebound* (1923), and has dramatized *The Great Gatsby* (1926), *The Good Earth* (1932), and *Ethan Frome* (1936).

DAVIS, Richard Harding (1864-1916), was born in Philadelphia, the son of Rebecca Harding Davis, pioneer naturalistic novelist and an associate editor of the *Tribune*. Davis became the most successful reporter of his time specializing in journalistic tours and war correspondence. In addition to his many travel books (*The West from a Car Window*, 1892; *Our English Cousins*, 1894, etc.) and military reports (*With Both Armies in South Africa*, 1900, etc.), he published 11 volumes of popular short stories, several novels (*The Bar Sinister*, 1903) and 25 plays.

DAVISON, Edward (1898-), was born in Scotland, served with the Royal Navy, and was a frequent contributor to London periodicals as well as the author of a book of poems before he entered Cambridge. After holding several editorial positions in London, he came to the U. S. in 1925, edited the *Wits Weekly* page of the *Saturday Review of Literature*, was a professor at Vassar, and lectured extensively. For many years he directed the Writers' Conference at the University of Colorado, and during World War II directed the army program to reeducate German prisoners of war. He is presently Director of the School of General Studies at Hunter College. His skillful, sensitive lyrics, although not in the present fashion, have attracted praise from many of the greatest English and American poets. His books include *Harvest of Youth* (1926), *The Ninth Witch* (1932), *Collected Poems* (1940), and *Some Modern Poets & Other Critical Essays* (1928).

DE FOREST, John William (1826-1906), was born in Connecticut, traveled extensively abroad, and received an honorary M.A. from Yale in recognition of his historical and travel books. He served as a Captain in the Union army, and used this experience as the basis for *Miss Ravenel's Conversion from Secession to Loyalty* (1867), the first realistic novel of the Civil War, and perhaps an influence on Howells. Mixed in with the triangular romance are authentic descriptions of the soldier's feelings in battle and vivid accounts of army corruption. His realistic study of South Carolina culture, *Kate Beaumont* (1872), was based on his work with the Freedmen's Bureau.

Deism was a religious philosophy of rationalism which rejected supernatural revelation and held that God rules natural phenomena by established laws perceptible to man. Its advocates included Voltaire and Rousseau in France and Jefferson and Franklin in America, although Franklin became pragmatically discouraged with the philosophy when he failed to collect from some Deists whom he had loaned money. Paine's *The Age of Reason* represents the clearest and most extreme exposition of the philosophy. Although it declined as a theological force in the early 19th century, it paved the way for such later reactions against Calvinism as the Unitarian movement.

DELL, Floyd (1887-), a prominent radical journalist and novelist, was the editor of *The Masses* and *The Liberator,* and wrote novels in revolt against convention, such as *Janet March* (1923).

DELMAR, Viña (1905-), wrote hard-

boiled novels of the jazz-age: *Bad Girl* (1928), *Kept Woman* (1929), and *Loose Ladies* (1929).

DERBY, George Horatio (1823-61), graduated from West Point in 1846, and for many years was assigned to the Southwest and California with the Topographical Engineers, commanding explorations into remote regions. In 1853, when a San Diego editor left Derby temporarily in charge of his paper, it was transformed into a hilarious burlesque, and Derby became known as a Western humorist. He published essays in several California papers under the pseudonyms of John Phoenix and Squibob, which friends later published as *Phoenixiana* (1855) and *Squibob Papers* (1865).

DERLETH, August [William] (1909-), has distinguished himself as one of America's most prolific writers, having published 56 books by his fortieth year and having 21 others ready for publication. He lives in a small Wisconsin town, Sauk City, the setting for most of his novels and poems.

DE VOTO, Bernard [Augustine] (1897-), was born and reared in Ogden, Utah, graduated from Harvard, and served as a lieutenant in the First World War. He taught English at Northwestern (1922-1927) and Harvard (1929-36). He edited *The Saturday Review of Literature* (1936-8), and has sat in the "Easy Chair" at Harper's since 1935. One of the most versatile writers of our time, he has achieved distinction as a critic, historian, and conservationist while contributing stories to popular magazines and writing eight novels under two names (John August is the pseudonym). Practically all of this production has centered on the Far West of his boyhood and wherever pertinent, has been spoken like a man holding a pair of Colt revolvers. *Mark Twain's America* (1932) set the record straight that Twain was a product of God and frontier in happy collaboration, and not, as a few critics had suggested, a Keats repressed by Puritan matriarchs. *Mark Twain in Eruption* (1940) presents his research into Clemens miscellany. He also wrote *Mark Twain at Work* (1942) and edited *The Portable Mark Twain* (1946). His monumental histories dissect the nation:*The Year of Decision: 1846* (1943), *Across the Wide Missouri* (1947), and *Course of Empire* (1953). Some of his two-gun essays have been collected as *Forays and Rebuttals* (1936) and *Minority Report* (1940). His novels include *We Accept with Pleasure* (1934) and *Mountain Time* (1947).

DE VRIES, Peter (1910-   ), grew up in Chicago and attended a Calvinist college in Michigan and Northwestern University. He was an associate editor of *Poetry* Magazine from 1938 to 1942, when he became a co-editor. In 1944 he joined the staff of *The New Yorker*, where many of his whimsical stories have appeared. His books include *But Who Wakes the Bugler* (1940), *The Handsome Heart* (1943), *Angels Can't Do Better* (1944), *No But I Saw the Movie* (1952), and *Tunnel of Love*

(1954). His rare humor usually turns on unusual applications of logic.

DEXTER, Timothy (1747-1806), wrote *A Pickle for the Knowing Ones* (1802) without any punctuation, although he invited his readers to "pepper and solt it as they plese." The book offers a variety of personal philosophies and observations, but Dexter is more interesting as the creator of his own legendary self. He manipulated his social status from farmhand to "Lord" Timothy of Newburyport, Mass., with a wheelbarrow vendor of haddock as his own "Poet Lauriet," and with a front-yard gallery of notables including his distinguished self. He made his initial fortune during the Revolution by collecting depreciated Continental currency, eventually stabilized by Hamilton. Thereafter his trading ventures included the sale of 42,000 warming pans to Cubans as cooking utensils, the shipment of crated cats to the Caribbean for rodent control, and the development of whalebone for corsets.

DICKINSON, Emily [Elizabeth] (1830-86), wrote more than 1500 poems, of which only 3 were published during her secluded life in Amherst, Massachusetts. Thus she was spared the tragic struggle with fame so common for those who write outside the fashions of their time. When, in 1862, she sent 4 poems to the kindly, conservative Colonel T. V. Higginson (q.v.), he attempted to straighten her into the paths of mediocrity. Fortunately, she abandoned hopes for recognition and continued to write her own "queer" poems on scraps of paper bag or envelopes, and tied them in bundles and hid them in bureau drawers for posterity. Emily was the daughter of a distinguished lawyer and legislator, Edward Dickinson, so dominating that he may have been half the God of her poems. Because her life was quiet and unheralded, scholars and pseudo-scholars have competed for the honor of identifying the lover she frequently mentions in her poems. The candidates include: Leonard Humphrey, her teacher (sponsored by Genevieve Taggard); Major Edward Hunt, the dashing spouse of Emily's close friend, Helen Hunt Jackson (sponsored by Josephine Pollitt); Benjamin Franklin Newton, law student in Edward Dickinson's office and an unconventional thinker who urged Emily to take herself seriously as a poet; and the Reverend Charles Wadsworth of Philadelphia, whom she met while enroute to visit her father in Washington, the subject of Martha Dickinson Bianchi's literary gossip and of Dorothy Gardiner's *Eastward in Eden.* Wadsworth is the only likely candidate, and then may qualify only if we conceive a Beatrice-Dante relationship, the mortal serving as the personification of a wish: "Love is anterior to life,/ Posterior to death,/ Initial of creation, and/ The exponent of breath." Where her love poems picture marriage or consummation, the scene is heaven; however, it is not Dante's *stil nuovist* heaven. She addresses her reader as if he were a fellow tourist: "Going to heaven! If you get there first,/ Save just a little space for me." When she is ready to depart, it will be as casual as if she were going to the Berkshires: "Because I could not stop for Death,/ He kindly stopped for me;/ The carriage held but just ourselves/ And immortality." These are samples of the magnificently conceived poems that Emily asked her sister, Lavinia, to burn after her death. Fortunately, Lavinia turned for advice to Emily's friend, Mabel Loomis Todd (q.v.), and she turned to Colonel Higginson. Together they selected 115 *Poems of Emily Dickinson* (1890). The book attracted so much attention that *Poems: Second Series* (1891) and *Poems: Third Series* (1896) were also published. A quarrel between the Todds and Lavinia over a few feet of land resulted in the transfer of the manuscript estate to Emily's niece, Emily Dickinson Bianchi, who edited *The Single Hound* (1914), *Further Poems* (1929), and *Unpublished Poems* (1936). Mrs. Todd left 650 unpublished poems to her daughter, Millicent Todd Bingham, who issued *Bolts of Melody* (1945). Because the manuscripts that Emily Dickinson left included fragments, notes, and variants of a single poem, each editor has needed to make his own decision as to what her final choice would have been. The miscellaneous publications have prevented any arrangement of the poems according to chronology or subject. Fortunately, Harvard University now has the combined manuscripts, and Thomas H. Johnson is in the process of editing a definitive edition of Miss Dickinson's complete poetry.

Dictionary of American English, edited by Sir William Craigie and James R. Hulbert, was compiled at the University of Chicago, and based on the Oxford Dictionary which Sir William had assisted in producing. It studies words which have a distinctive, American meaning, whether as the result of coinage or because of different developments since their importation from England. Four volumes were published between 1936 and 1944.

DIXON, Thomas (1864-1946), author of more than three dozen books and plays, depicted the Reconstruction South strictly from the Southern point of view. *The Leopard's Spots* (1902) and *The Clansman* (1905) presented Dixon's inflammatory defense of white supremacy and the Ku Klux Klan. His other books were polemics on race relations, socialism, and women's rights. He also wrote the motion picture, *The Birth of a Nation* (1915), based on *The Clansman*. Born near Shelby, North Carolina, Dixon graduated from Wake Forest College, later was a lawyer, politician, minister, and lecturer. Widely read in his day, his books now seem rather like bitter period pieces.

DOBBIE, J[ames] Frank (1888-), secretary of the Texas Folk-Lore Society, former professor at the University of Texas and at Cambridge, has written several books on Southwestern folklore including *Coronado's Children* (1931) and *Guide to Life and Literature of the Southwest* (1943).

DOOLITTLE, Hilda (1886-), known also as H. D., was born in Bethlehem, Pennsylvania, where her father was a mathematics professor at Lehigh University. Poor health forced her to leave Bryn Mawr after two years. On a trip to Europe in 1911 she met Ezra Pound and became one of the first Imagists. She married Richard Aldington, settled in England, and took her husband's position as editor of the *Egoist* while he served in the war. Her first volume of poems, *Sea Garden* (1916), shows her experimentation with objective conceits. In *Hymen* (1921), her imagery is better condensed to carry intense emotion. Her *Collected Poems* were published in 1925, although several volumes have appeared since which continue her Imagistic concentration in dealing mainly with classical subjects from the modern point of view.

DOS PASSOS, John [Roderigo] (1896-), was born in Chicago, traveled extensively with his family as a boy, and graduated from Harvard in 1916. He then went to Spain to study architecture, but soon joined a French Ambulance Service, served with the Red Cross in Italy, and later became a private with the U. S. medical corps. In Spain after the war, he wrote *Three Soldiers* (1921), which aroused a heated controversy over the charge that it insulted the army. He achieved his reputation with *Manhattan Transfer* (1925), a kaleidoscope of episodes and characters adding up to a horizontal cross-section of New York City. *The 42nd Parallel* (1930), *1919* (1932) and *The Big Money* (1936) were collected as the trilogy, *U. S. A.*

STEPHEN CRANE

EMILY DICKINSON

T. S. ELIOT

Photo: Lilyan Lowmans-Chauvin

WILLIAM FAULKNER

(1938), relating the first three decades of this century with a Newsreel technique and demonstrating the decay of our civilization due to commercial exploitations. He has been highly experimental in writing many additional novels of immediate but perhaps not lasting interest.

Double Standard; a term used to define the dichotomy between the moral behavior expected of men and the much more stringent restrictions placed upon women. Among certain levels of society, particularly during the 19th century, it was expected that a husband might violate the marriage code, whereas his wife would be chastised if she so much as lunched with another man. A violation of the double standard was considered to be as offensive in literature as in life.

DOWNING, Major Jack, pseudonym of Seba Smith (q.v.)

DRAKE, Joseph Rodman (1795-1820), a New York druggist, wrote satirical verses with his friend, Fitz-Greene Halleck, which became known as the "Croaker Papers." They were sent anonymously to the New York newspapers in 1819, and an unauthorized edition was published soon afterward. After Drake's early death from consumption, Halleck wrote the elegy, "On the Death of Joseph Rodman Drake," which is frequently reprinted in anthologies. Drake's own work was published as *The Culprit Fay and Other Poems* (1835).

DRAMA IN AMERICA:

Colonial Drama: The English theater suffered a veritable eclipse as the result of the hostility of the Puritans during the 17th century, and the Puritan theocracy of New England could hardly be expected to show a greater tolerance for drama. If literature as a whole was suspect, the stage was considered to be the greatest evil ever invented by Satan to ensure the damnation of sinners. Even in Virginia the immigration of English actors was forbidden in 1610, and 55 years later some young men were brought to court for "acting a play." But the Restoration of the English Crown (1660) led to a relaxation in the South, and in 1703 a play was presented in Charleston by a company of professional actors. There is also a record of Richard Hunter being granted a license to produce plays in New York in 1700, but no record of the production. During the next 50 years, companies of strolling players performed in coffee houses or barns along the Atlantic seaboard. The tragedy of *Cato* was produced in Philadelphia (1749), *Richard III* was presented in New York (1750), and *The Merchant of Venice* was produced at Williamsburg (1752) by the "American Company" which included members of the famous Hallam family. Theaters were built in Philadelphia (1766), New York (1767), Annapolis (1771) and Charleston (1773). As members of society began to appear at the performances, including elegant ladies, and finally when General Washington became a frequent visitor to the theaters, hostility declined everywhere in the colonies except in Boston.

The first play written by a native American and produced professionally was a tragedy in blank verse concerning a mythical Oriental, *The Prince of Parthia* (1767) by Thomas Godfrey, based on Elizabethan theatrical conventions. The Revolutionary War stimulated a new interest in drama as plays were written in support of each side. Mrs. Mercy Warren wrote a satire on the British, *The Adulateur* (1773), and was perhaps the author also of *The Block-heads*. *The Battle of Brooklyn* (anon., 1776), which lampooned Washington, and *The Blockade* (1776) supported the Tory position. After the Revolution it was natural that American patriotism should continue to dominate the theater. Royall Tyler's *The Contrast* (New York, 1787) places a pure American heroine in jeopardy from an English cad, and the prologue is significant: "Why should our thoughts to distant countries roam/ When each refinement may be found at home?" James Nelson Barker continued this trend with *Tears and Smiles* (1807), but he substituted the French for the English in the role of villain. The fact that another of his plays caused a riot when it was produced in Philadelphia shows the theater functioning as a social commentary, for although *The Embargo* (1808) manuscript was lost, it is known to have supported the administration and to have alienated the merchants. Barker was also the first American to exploit the Indian theme, in his play about John Smith and *The Indian Princess* (1808). Again, he was the first of many dramatists to use the Salem witchcraft trials as a basis for drama, in *Superstition* (1824).

One of the first major producers and "men of the theater" was William Dunlap, who wrote and adapted more than 60 plays. Although he did not achieve great stature as a dramatist, he advanced the cause of the theater in America by acting as a professional producer with a constant and varied set of plays to present, from Shakespeare to his own creations. Furthermore, he took the position that the state of the national theater was a measure of the nation's refinement and manners. He was an able producer with high ideals who conscientiously studied the problems of drama and production. He was not an extremist in advocating a theater of nationalism, feeling that the theater was more important to America than Americanism was to the theater.

Nineteenth Century: During the first half of the 19th century the main theme of the drama, in addition to nationalism, was escapism to romanticized and exotic history. N. P. Willis wrote about 14th-century Milan in *Bianca Visconti; or the Heart Overtasked* (1837), and his *Tortesa the Usurer* (1839) was set in Florence. John Howard Payne and Washington Irving adapted *Charles the Second; or The Merry Monarch* (1824) from the French play by Alexandre Duval. Robert Montgomery Bird used classical Rome as his setting in *The Gladiator* (1831), and his *Broker of Bogota* (1834) was about colonial South America. These historical plays were in the Elizabethan tradition of the English theater, and quite natural, therefore, to the American stage. A later English tradition was mirrored in Mrs. Anna Cora Mowatt's *Fashion* (1845), an American comedy of manners. Many historical plays used an American setting.

In addition to Barker's *Superstition,* there was *Witchcraft* (1846) by Cornelius Mathews, who also wrote *Jacob Leisler, or New York in 1690* (1848). Elizabeth Oakes Smith wrote about *Old New York, or Democracy in 1689* (1853). The prolific writer in every branch of literature, James Kirke Paulding, introduced a Cooper motif to the stage with *The Lion of the West* (1831). Thus nationalism continued to dominate the stage, and even the historical romances usually had a democratic theme.

If the theme of these plays was nationalism, their format was English or French, for America had not yet developed a national drama. The American theater, even more than other branches of literature, was dominated by Europe. Financial domination was the most limiting factor: there was no international copyright agreement until 1891, and European plays could be produced in America with no royalty cost, so that American playwrights worked under a great economic disadvantage in addition to the more natural disadvantages of competing with English prestige and cultural advantages. If a fiction writer, such as Hawthorne, could surmount this disadvantage by having his books published with private funds, or if Whitman could print his own editions of *Leaves of Grass,* it would be a fairly wealthy dramatist who could afford to produce his own plays. The early plays were often competent, but they fell quite short of distinction. Even the American copyright laws were inadequate before 1865; a play could be pirated within the United States, and

the dramatist was protected only if he was also a professional producer or actor. Facing these problems, the Americans who did manage to achieve some measure of success were compelled by competition to write competent plays, but if the various pressures eliminated feeble vehicles, they also worked against the development of a powerful national drama. Great actors appeared on the scene—Edwin Booth, Charlotte Cushman, E. L. Davenport, Edwin Forest, Clara Morris, James E. Murdoch, and Lester Wallack—but the great plays they presented had been written by Shakespeare. The American plays they presented were often several decades old, and free of royalty obligations.

It was prophetic that the earliest dramatist to achieve some measure of greatness should abandon nationalism and concern himself with humanity. George Henry Boker produced 11 plays while he was still a young man, but each of his fine tragedies in blank verse ran less than two weeks (*Calaynos,* however, was pirated by a London producer and merited a hundred performances there). His best play, *Francesca da Rimini* (1855), utilizes a sympathy for the deformed husband, Lanciotto, to produce a classic theme of destiny, with destiny the villain. Lanciotto loves both Paulo and his wife, Francesca; indeed, had sent Paulo to woo Francesca in his place. In each of Boker's plays, the tragedy turned on a militant assertion of human dignity. Boker was himself a victim of the same irony of fate that he so aptly exploited in his plays, for, many years after he had given up drama, completely discouraged by the apathy to his work, traveling companies began to use his plays, and *Francesca da Rimini* became a standard repertory drama for Lawrence Barrett in 1882.

The traveling company was a post-Civil War development that took advantage of the growing rail transportation to exploit audience potentials of inland America. The plays presented might be Shakespeare, or they might be domestic comedies concerning the shanty Irish immigrant or the country bumpkin. A multitude of plays was turned out by "theater people," but this was mass production of entertainment rather than creative art. Dion Boucicault transferred French plays to an Irish setting, and there was considerable truth in the definition Ambrose Bierce gave to the word "dramatist" for *The Devil's Dictionary*: "One who adapts plays from the French." The adaptations, however the original plays might be altered in the process, at least provided better fare than the average native production, which was stock melodrama with black-and-white delineations of vice and virtue. The stage, after the Civil War, was comparable to the motion picture industry after World War I, and the lack of quality derived more from the condition of the audiences than from a dearth of creative ability. In each development, producers were concerned with financial remuneration rather than art, and aimed their products at the lowest quotient of national taste, which would provide the largest audience. The sentimental tradition of the theater continued until it was taken over by Hollywood, and in its

later years was represented, at its best, by James A. Herne's *Hearts of Oak* (1879), *Way Down East* (1898) by Lottie Blair Parker, Winchell Smith's *Lightnin'* (1918), and Anne Nichols' *Abie's Irish Rose* (1922). From Hollywood, the tradition was inherited by radio and television, and continues at full force in the contemporary "soap opera."

Herne had grown up in a theater family and was nursed on such sentimental plays as *Uncle Tom's Cabin* and *East Lynn*. When he wrote *Hearts of Oak* with a young San Francisco lad named David Belasco, it was natural that he should begin in the sentimental tradition. Later he tempered sentimentality with realism as he came under the influence of Hamlin Garland and William Dean Howells, who introduced him to the works of Ibsen and Tolstoy. *Drifting Apart* (1888), *Margaret Fleming* (1890), and *Griffith Davenport* (1899) were praised by such critics as Garland and William Archer, but they were financial failures. Herne was saved, financially, by the success of *Shore Acres* (1892), which used stock situations to carry a mild dose of social commentary and realism. His career demonstrated that the audiences needed to be led to greater drama, gradually, and he is remembered as a pioneer of the modern drama. Another pioneer was Bronson Howard, who also began his career by writing in accepted patterns, and matured to write social comedies in the Henry James pattern as well as the Civil War play, *Shenandoah* (1888).

The modern theater was not to arrive, for all the pioneering of such men as

Herne and Howard, for twenty years after the turn of the century. In the interim, America continued to enjoy a very competent theater under the leadership of such men as David Belasco, Augustus Thomas, and Clyde Fitch. The battle of realism was being waged primarily in the field of the novel, and the theater would need to wait until the forces of romanticism were in retreat. Dreiser's *Sister Carrie* could wait on the shelf twelve years, after an abortive publication, for the tolerance of publishers and public. A play production can not wait for an audience for more than a few nights. The economic factors of play production make the professional theater conservative. Meanwhile, the modern theater was developing in continental Europe under the leadership of such powerful dramatists as Strindberg, Hauptmann, Gorky, Chekhov, Schnitzler, Maeterlinck, D'Annunzio, and Shaw.

In the midst of the commercial theater appeared a University of Chicago professor who aspired to rival Shakespeare. William Vaughn Moody's trilogy, *Masque of Judgment,* was never produced professionally, and he made no concessions when he wrote it that would permit its production in the theater of 1900. The three poetic dramas, *Masque of Judgment* (the first, giving its name to the trilogy), *The Fire Bringer,* and *The Death of Eve* (unfinished) are in the tradition of Hawthorne and Melville, and they are intrinsically American in their manipulation of two themes: (1) Calvinism with its tenet of predestination and eternal sin; (2) humanitarianism and reason examining the problem of sin. This combination is also reminiscent of Milton, and Moody wrote with a Miltonic sweep and courage, if not quite so well. Thus, there were dramatists on the scene who aspired to produce a great American drama. Moody's friend, Edwin Arlington Robinson, also tried his fortune in drama, and received even less encouragement. A play which does not reach production can not, perhaps, be reckoned as a play. The great dramatists of the past have needed to compromise to some degree with the demands of their public. The degree and kind of compromise which an audience demands may shape the theater. It may be argued that Moody and Robinson were at fault in looking backward to classical models rather than anticipating the demands of the modern theater. Whatever the argument, it must be recognized that the desire to produce quality existed, and that the encouragement of quality did not. Moody recognized that compromise was necessary, and the success he achieved with *The Great Divide* (1909) reveals that he was capable of producing a play for his place and time. The degree and kind of his compromise may reveal, in turn, something about his audience. Plays about the West were in vogue: Thomas' *In Mizzoura* (1893), Fitch's *The Cowboy and the Lady* (1899), and Belasco's *The Girl of the Golden West* (1905) had successfully exploited a new American romanticism which imposed heroic proportions on the pioneer or the cattleman. Moody fitted his study of sin to this pattern in his portrayal of the great divide between Puritan conscience, represented by the heroine, and frontier abandon as demonstrated by the hero. The curtain falls on their quite necessary reconciliation. Moody compromised less in *The Faith Healer* (1909), which was a better play, and a failure.

The Modern Theater: The problem of securing a modern American drama was to be solved, not by compromise with the American audience, but rather by leading the audiences to a compromise with new mores and dramaturgical techniques. Part of this task was accomplished for the theater by the novelists. The device which would complete the transition was a new institution, now called "the little theater" movement. During the summer of 1915 a group of Bohemians from Greenwich Village, New York, were vacationing at Provincetown, Massachusetts, and to amuse themselves they staged some one-act plays in a deserted fish house. They amused themselves so well that they continued the project during the following summer, and produced a one-act melodrama, *Bound East for Cardiff*, by a young man living in Provincetown named Eugene O'Neill. The following November they presented this and other one-act plays in New York City at the small Provincetown Playhouse on Macdougal Street. The meeting of these intellectuals, whose interest in the theater was purely amateur, with O'Neill, the son of a professional actor in search of new horizons, was a fruitful serendipity for the modern theater in America. The little theater movement had originated in Paris with the founding of Antoine's Théâtre Libre (1887), and continued

with the establishment of Stanislavsky's Moscow Art Theater (1890), the English Independent Theatre (1891), and the Irish Literary Theatre (1899). It was slow in coming to America, but it expanded rapidly, once under way. Another little theater group, the Washington Square Players, founded in 1915, was reorganized as The Theatre Guild in 1918, and was so successful in presenting the work of Shaw that it was able to build its own million-dollar theater uptown by 1925. George Pierce Baker's 47 Workshop at Harvard and Frederick H. Koch's Carolina Playmakers at Chapel Hill were academic branches of the same movement. The Hull House Players in Chicago and the Neighborhood Playhouse in New York were operated by social settlement organizations. By 1918 there were more than 50 little theaters in the country.

The only major dramatist "discovered" by the little theaters in their early period was Eugene O'Neill, who probably would have been discovered by commercial theaters in time. What the little theaters did discover and magnify was an audience for experimental drama. When the Theatre Guild showed the professional producers that money could be made from the presentation of Strindberg, Molnár, Shaw, and O'Neill, or even from *The Adding Machine* (1923) by a relatively unknown American playwright named Elmer Rice, then Broadway became more liberal in its presentations, and the American professional theater entered a new period of mature dramatic production. Eugene O'Neill was the pioneer American dramatist for the new development, and the Province-town group was the pioneer in breaking a path for him. The Theatre Guild, gifted with shrewd management, pioneered the path from Greenwich Village to Broadway.

O'Neill was an especially able pioneer because of the variety of his dramatic interests and his restless experimentation. The path he broke was exceedingly broad. His early one-act plays, such as *Bound East for Cardiff*, were poetic and romantic in tone, and did not break too abruptly, therefore, with established traditions, but the realistic dialogue broke with the traditional clichés of romantic drama. *Diff'rent* (1920) was influenced by Freudian psychology. *Emperor Jones* (1920) treated a sociological problem with mysticism and poetry, and broke with the sociological clichés. *Beyond the Horizon* (1920) and *Anna Christie* (1921) were examples of fairly simple realism. *The Hairy Ape* (1922) is realistic on the surface, but O'Neill counterpointed the realism with a cacophony of Strindbergian Expressionism (q.v.). This play appears to develop a sociological message, on surface reading, but further study reveals that O'Neill was exploiting for the first time what would become his persisting theme, a philosophy of despair over man's frustration in trying to identify his place and and function in society. *All God's Chillun Got Wings* (1924) again appears on its surface to be a sociological problem study, but again O'Neill revealed his literary stature by providing apprehensions, which are more profound than solutions.

The professional theater had begun to show unusual if not great courage as early as 1916, when the first of Clare Kummer's sophisticated comedies, *Good Gracious, . Annabelle!,* was produced. Realism came to Broadway in 1920 with *Miss Lulu Bett,* the dramatization of Zona Gale's novel. Folk drama appeared the following year: *Wake Up, Jonathan,* by Hatcher Hughes and Elmer Rice, presented Southern mountaineers. The definite change in the professional theater is best marked, however, by the 1924-5 season, when two plays by relatively unknown authors appeared, presented as commercial ventures without a pioneering introduction, and characterized by a complete break with previous mores and sentimentality. Neither play could have been produced by the professional theater in 1916. The plays were: *What Price Glory?,* by Laurence Stallings and Maxwell Anderson, produced by Arthur Hopkins of the commercial theater; and *They Knew What They Wanted* by Sydney Howard, produced by the Theatre Guild, which was then operating as a professional theater. Their immediate success demonstrated that the period of pioneering was over, that these plays reflected the convictions and emotions of their audience, and that the American theater would now be able to provide an artistic version of public attitudes rather than the clichés of the past. They were also shocking, but the public was ready and even eager to be shocked.

The liberation of the theater for an intellectual approach to moral and social problems seems to involve also the liberation of moral and social attitudes concerning taboos of expression and thought. *What Price Glory?* displayed an intellectual cynicism concerning the glory of fighting a country's battles, and it also displayed a ribald, almost blasphemous picture of the language and life of men who had been called to fight these battles. *They Knew What They Wanted* presented Tony, a middle-aged California winegrower of Italian ancestry, eager to marry a waitress who bore another man's illegitimate child. It is significant that this was not a thesis play, that Howard felt no need to defend the unwed mother. If the play had a thesis, it was "Know what you want, ensure that the want is sensible, and everything will come out all right." Had Tony reacted to his situation with the blind passion of Othello, the play would have been a tragedy. Older mores would have demanded some such action, tragic, melodramatic, or comic, to maintain the mores of respectability. The new theater joined the novel in questioning the ethics of respectability and in substituting humanism. This humanism revolved around human rather than sociological issues during the twenties, when Naturalism, even in the novel, was temporarily discarded for optimism concerning the economic potential of the American way of life.

Traditionally, comedy may be expected to emphasize the intellectual, since it has been defined as the story of those who think, as opposed to tragedy, which depicts those who feel. This traditional definition of comedy had been lost in America to sentimental slapstick and vaudeville. S. N. Behrman's *The Second Man* (1927) helped to restore the intellectual quality to comedy, for when the hero is confronted with a rival in love, bent on avenging the honor of the girl, he merely invites the rival to sit down and talk the matter over, without heroics. "Common sense" on the other hand, can have no place in tragedy, according to traditional definitions, and the production of tragedy in a country which worships common sense was consequently limited. The two major explorers of modern tragedy were Eugene O'Neill and Maxwell Anderson, although they approached the problem with a difference that has been defined by the observation that O'Neill was finding words for passion, whereas Anderson was finding passion for words. The distinction is apt, but incomplete. O'Neill was a restless experimenter and innovator; Anderson actually worked more in the tradition of George Henry Boker than in the development of the modern theater. After delineating the modern theater, in collaboration with Stallings (*What Price Glory?*), and trying comedy, he went back to such historical subjects as *Elizabeth the Queen* (1930), *Mary of Scotland* (1933), and *Valley Forge* (1934), writing traditional drama in a loose blank verse. Later he used the same formula for contemporary problem plays, *Winterset* (1935), and *Key Largo* (1939), and he was successful in giving the first of these a poetic quality which keeps to the direction of tragedy. It has often been argued that classical tragedy can not exist in a country and culture that has no universal belief or tradition, and finding a way to circumvent what Eliot calls "the dissociation of modern sensibility," has been the concern of modern dramatists who consciously set out to write tragedy, from Anderson to Arthur Miller. The question arises, can a modern audience, burdened with disbelief, view *King Lear* as a tragedy? If the answer is yes, that *Lear* becomes a tragedy because we accept the tradition of Shakespeare's period and suspend disbelief, then the answer to modern tragedy might be the return which Anderson made to an Elizabethan setting and situation. The question of whether *Winterset*, which uses a contemporary setting and a sociological situation, can be tragedy received an unexpected answer when the play was produced as a motion picture, for the delicate sensibility of the cinema audience was protected by a device which saved the hero from certain death. Those who feel that this device violated the whole dramatic direction of the play will be apt to conclude that *Winterset* achieved genuine tragic dimensions. Those who feel that the change produced no distortion, that the play could function with either a happy or an unhappy ending, are bound to conclude that the unhappy ending was not sufficient in itself to produce genuine tragedy.

Whatever his actual intentions may have been, O'Neill did not appear to be struggling with the problem of producing tragedy, of fitting a play to a definition, and some of his plays come much closer than Anderson's to our sense of what tragedy must be, however we define that sense. Joseph Wood Krutch conceives of tragedy as the play which expresses and conveys the conviction that to be a man is a great and terrible privilege, an art which consoles man in defeat by reassuring him of his nobility. O'Neill faced the problem of producing modern tragedy directly when he transferred the Greek story of Electra to an American reality and setting. *Mourning Becomes Electra* (1931), a trilogy of three plays for three evenings, is closer to Aeschylus and Euripides than anyone would expect that a play about the Civil War could possibly be: the story is almost identical to the Greek; yet the characters and even the motives are modernized to a rational world without a set of gods. Perhaps no modern play comes closer to the classic definition of tragedy. Its failure to come all the way might derive, not from the lack of gods, but from the lack of eloquence, for O'Neill did not even attempt to write poetry. (David Hume [1711-1776], in exploring the question of why we enjoy the displeasure of tragedy, found eloquence to be the conditioning factor.) In *Desire Under the Elms* (1924) O'Neill employed a Calvinist God for supernatural overtones, and in *Strange Interlude* (1928) he used Freudian psychology to achieve super sensual overtones. In *Desire* the god added nothing to the tragic effect, for neither O'Neill nor his audience believed in him. Psychology does no more for *Strange Interlude*, even though psychology is believed, for it reduces man (or in this case, woman) to a clinical study. The question of whether tragedy can be achieved in the modern theater may be entirely irrelevant to any appraisal of the modern theater, for the greater approximation of the *Electra* play to tragedy does not make it, in proportion to this approximation, a greater play than *Desire Under the Elms*. Certainly there seems to be no question that O'Neill was the greatest force in modern American drama, as well as the most vital explorer of modern dramaturgical possibilities.

There has been no problem of comedy becoming reconciled to the modern mood. Since American literature had a long tradition of humor, it was natural that the new comedy should be of a native quality. Robert Sherwood showed a Shavian influence in writing his pseudo-historical comedy about *The Road to Rome* (1927), and his *Reunion in Vienna* (1931) captures the continental comic style, but the majority of the comic dramatists used American scenes and motivation. The difference in the new comedy, as noted earlier, was seen in its closer adherence to the traditional definition of comedy as distinguished from the sentimental melodramas, decorated with laughs, which had previously dominated the American stage. European examples, from Wilde to Schnitzler, may have helped to set the new pattern. It is very likely, however, that the new audience for sophisticated comedy was conditioned to a large degree by the new national appetite for cynicism and reason explicit in the success of *What Price Glory?* Again, the importance of the audience can easily be underestimated. In 1906 the sentimental *Brewster's Millions* (by Winchell Smith) was far more

successful than Langdon Mitchell's more sophisticated *The New York Idea,* and Jesse Lynch Williams' Shavian *Why Marry?,* although it was awarded the Pulitzer Prize in 1918, did not touch the popularity of Smith's *Lightnin'* (1918). The sentimental tradition continues to attract large audiences in New York, especially in musical comedies, from the long series of Rodgers and Hammerstein productions to the 1953 Borodin version of *Kismet,* based on Edward Knoblock's sentimental drama of forty years earlier. But there is also a large and appreciative audience for genuine comedy. The major distinction between the two types, real and sentimental comedy, is that the root idea, the theme of the true comedy is itself comic, an enlarged epigram or witticism, and not a sentimental love story. Professor Krutch defines comedy as the kiss given to common sense behind the back of respectability. Comedy is intellectual. Tragedy is emotional. The sentimental drama mixes the laughs of comedy with the emotions of tragedy, and the result is a conglomeration calculated to confuse the audience into believing it has been entertained. Traditional definitions make a clear distinction between tragedy and comedy that allows no mixture of the two. The technique which develops laughter into comic drama, and which had been missing in most American plays, may be defined as "wit," the intellectual manipulation of ideas into incongruities which are pleasing because they relieve—rather than heighten—our emotional tensions. Mere common sense may be witty if it is unusual—a phenomenon that was often the basis of Mark Twain's humor. The scene in Behrman's *The Second Man* in which the hero reasons with his intended murderer provoked laughter because audiences had been conditioned to think of honor in heroic clichés. There is nothing essentially funny about the lines when they are taken out of their context.

Philip Barry had been producing rather whimsical plays before *The Second Man* was produced in 1927, but in 1928 he too exploited wit to produce a drawing-room comedy, *Holiday,* followed by other genuine comedies: *The Animal Kingdom* (1932), and *Bright Star* (1935). *The Philadelphia Story* (1939), although it deviates into romanticism, also owes much of its charm to Barry's use of wit. Meanwhile, Behrman continued to produce such excellent comedies as *Brief Moment* (1931) and *Biography* (1932), only to dilute his later plays with the poison of ideology under the pressures of the economic depression and the threat of Fascism.

These two pressures were a considerable threat to the new American drama which had experienced such an auspicious beginning. Almost every dramatist of the twenties, with O'Neill remaining a conspicuous exception, succumbed to ideology in the thirties, and the new dramatists were born to it. Messages tend to destroy drama because they are essentially sentimental; furthermore, if the argument is convincing, if the audience is persuaded, the play has no further utility. Of the left-wing dramatists who arose in social protest during the thirties, only Clifford Odets suc-

ceeded in overcoming the liabilities of social protest, and then only in two of his plays, *Awake and Sing* (1935) and *Rocket to the Moon* (1938). Odets succeeded in spite of his convictions because his sense of drama and his perception of people as human beings were sufficiently strong to overcome his didacticism. His dramaturgical power is also evident in the more direct message-play, *Waiting for Lefty* (1935), produced by the Group Theatre (a leftish extension of the Theatre Guild), but the play is good propaganda and "good theater" rather than good drama.

Ideology subsided as the threat of Fascism developed into the actuality of war, which in turn relieved the economic stalemate of the depression. A new theme, which had appeared sporadically during the previous 25 years, began to gain in popularity. The practical attitudes of the American public have spoiled its appetite for fantasy throughout most of our literary history, and few novels or stories have succeeded in presenting an imagined reality. The visual reality of the stage or screen, however, makes fantasy more believable, especially if it is presented in a realistic context. O'Neill had used a special type of fantasy in his expressionistic plays, and Anderson employed it in *High Tor* and *The Star Wagon* (1937). Marc Connelly's *The Green Pastures* (1930) was a Negro fantasy of heaven. The greater trend toward fantasy began with the production of Thornton Wilder's allegory of *Our Town* and Paul Osborn's *On Borrowed Time* in 1938. Wilder, who used the fantasy of expressionism, developed his technique even more fully in *The Skin of Our Teeth* (1942). William Saroyan's *My Heart's in the Highlands* and *The Time of Your Life* (1939) also deserve to be recognized as part of this trend, for although they present no supernatural incidents, they have the lightness, whimsy, and dreamlike quality of fantasy. Comedy and fantasy were aptly blended in Mary Chase's *Harvey* (1944), although she violated the true comic spirit in maintaining a secondary plot of sentimental love. The main plot, however, of an inebriate and his six-foot rabbit, is in the Behrman tradition of common sense evoking humor because of its rarity. Mrs. Chase continued the development of fantasy in *Mrs. McThing* (1951).

A touch of Chekhov, whether an influence or a coincidence, was visible in the best works of Odets; however, the serious continental drama appeared to move too slowly to develop a more definite influence on the American theater until *The Glass Menagerie* was produced in 1944. Tennessee Williams displayed a unique ability to create a delicate and vital drama from very ordinary experience through the intense development of character, instead of from the normal proletarian devices of comic dialect, shocking crudity, or didactic pathos. He was also amazingly successful in discarding the distinction between comedy and tragedy without producing complete sentimentality. The extremes of tragic situation are even more strongly contrasted with ribald humor in *A Streetcar Named Desire* (1947). Neither play develops the heroic proportion essential to classic tragedy, and perhaps the tragic element could better be described as a gentle satire on human, especially female, frailty. Williams' plays may not be great, but they have a quality of poetry which lifts them above the average mixture of the comic and sad.

Experimentation has been predominant among recent plays in their staging rather than in the narrative structure. *The Glass Menagerie* achieved the effect of three scene shifts with a single set by the use of lighting to emphasize one and block off another area of the set. The same device was used in Arthur Miller's *The Death of a Salesman* (1949) —another attempt to solve the problem of modern tragedy, and an attempt which discards classical definitions. Truman Capote's *The Grass Harp* (1951) was a successful experiment in the use of the circular theater, with the audience surrounding the stage and a minimum use of "props." This general change in the direction of experimentation was caused by the new economic exigencies that followed the Second World War. Labor costs rose considerably, and labor unions not only enforced higher wages for craftsmen who serve the theater, but also resisted amateur production which would be competitive to the commercial theater. Their policy has forced the contemporary theater toward a closer observance of the Aristotelian unities. It became economically desirable to build a single set, but the use of lighting allowed a measure of scene variety within a single location. Even the more commercial dramatists have needed to call upon an unusual dramaturgical skill to limit production costs, a task that was accomplished to perfection by Elliott Nugent in *The Voice of the Turtle* (1945), which used a single set with a cast limited to one man and two girls. The problem of developing a post-war little-theater movement under union restrictions has recently been solved by compromise. Founders of the Phoenix Theater (1953-   ) agreed to submit to union regulations if any of their experimental plays becomes a commercial success.

In summary, the fine distinctions between the tragic and comic spirits which characterized the new theater of the twenties seem to have disappeared by mid-century. The extraordinary number of fine playwrights who appeared after the First World War has not been replaced by a comparable group after the last war. The fact that so many of the earlier group emerged from George Pierce Baker's Drama Workshop (first at Harvard, later at Yale) suggests that he alone may have been a greater force in the theater's destiny than is generally recognized, granting the recognition he has received as a teacher. The small group of recent dramatists reveals an eagerness to produce serious drama rather than sentimental melodrama, and there is a quality of poetry in their work which may be a sustaining element. The younger playwrights may have evolved a successful human comedy that depends, not on wit, but on a pleasing revelation of the incongruities of life, and a sympathetic revelation of the vanity of human desires. Perhaps the final question of their quality will depend upon how well the pathos of their situations keeps its strength. Meanwhile, the emergence of T. S. Eliot as a commercial success with *The Cocktail Party* (1950) and *The Confidential Clerk* (1954) indicates that the audience for sophisticated drama is still alive.

Popular narrative has shifted its emphasis from fiction to drama with the rapid development of television. This competition may be partly responsible for the production of an increasing number of fine motion pictures. It is difficult to decide the degree to which these dramatic mediums have siphoned potential playwrights of quality from the legitimate theater, but there is no question that they offer far greater remuneration. The increased cost of production has again made the theater exceedingly conservative, and the question arises, if a new O'Neill should happen to be writing plays today, is there much chance that his plays would reach production? Audiences in New York are rabid in their desire to buy tickets for "the best" plays, but paradoxically, if tickets become immediately available, the playgoers lose all interest in buying them. In short, the American theater seems presently to be in a situation where anything can happen, and probably will.

DREISER, Theodore [Herman Albert] (1871-1945), was born in Indiana, the 12th child of an extremely poor and religious family. He attended Catholic schools in Terre Haute, Sullivan, and Evansville, and a public school in Warsaw, becoming an avid reader of the great American and English novelists.

A friend financed his attendance at Indiana University for one year, the limit of Dreiser's endurance for higher education. He held several jobs in Chicago before advancing to journalism, which he also pursued in St. Louis, Pittsburgh, and New York. During his twenties he wrote poems in the Whitman manner (published as *Moods, Cadenced and Declaimed,* 1926) and sold feature articles to the magazines. His first novel, *Sister Carrie* (1900), attracted the attention of Frank Norris, then a reader for Doubleday. After about 2,000 copies were sold and Norris had sent a hundred to reviewers, the publisher's wife was so shocked, on reading the book, that she insisted the edition be withdrawn. The initial situation in *Sister Carrie* is stock melodrama: a small town girl meets a traveling salesman on her way to Chicago where the sweat-shops are so intolerable that she becomes his mistress. But it is the girl who deserts the salesman. She finds a more sophisticated lover in the socially established Mr. Hurstwood, with whom she elopes. In New York, Carrie rises to theatrical stardom while Hurstwood degenerates, finally to a grave in potters field. For Carrie the wages of sin are increasing success. Dreiser had violated the mores of both "the double standard" and "poetic justice" (qq.v.), and this twofold heresy provoked a storm of protest. Now no editor would touch his inoffensive articles, not even William Dean Howells, the ostensible champion of realism. In the following decade Dreiser found editorial work with pulp and women's fashion magazines until *Jennie Gerhardt*

(1911) achieved sufficient success to establish him as a professional novelist. Even the offensive *Carrie* was reissued. America did not yet approve of Dreiser, but he could be tolerated. *Jennie Gerhardt* expands and improves the theme of his first novel, and in the end, Jennie sacrifices her own interests for those of her second lover. With *The Financier* (1912), Dreiser turned his attack more specifically upon the economic mores, portraying the Machiavellian rise to power of the unscrupulous Cowperwood, whose career he extended to a trilogy of desire with *The Titan* (1914) and *The Stoic* (1948). *The Genius* (1915) is a portrait of the artist as a weak man. *An American Tragedy* (1925) his first financial success, has the most unified structure of his novels, but contains no strength to compensate for the weakness of its protagonist. Clyde Griffiths' great opportunity to achieve social respectability is threatened by the pregnancy of a girl he has abandoned. He selects murder as the solution, but at the fatal moment he is too weak, and she is killed, not by his act of will, but rather by the circumstances he had set up for accomplishing the deed. The situation is perhaps the most subtle delineation of drama in all of Dreiser's works. He has been criticized for his heavy-handed manipulation of theme, his clumsy style, and his occasional verbosity. He has been praised for his dynamic perception of the gulf between our ideals and our provisions for their realization. No other writer of our time has faced such a barrage of criticism or misunderstanding, and no other writer in

American letters has so often been acknowledged as the pioneer of the contemporary novel.

DUNBAR, Paul Laurence (1872-1906), used his native Negro dialect and folklore in producing *Lyrics of Lowly Life* (1896), *Lyrics of the Hearthside* (1899), and *Lyrics of Love and Laughter* (1903). Of his four novels, *The Sport of the Gods* (1902) is considered to be the best. He is one of the few Negroes who have achieved recognition from their own as well as the white race, but he was disappointed to find that his dialect verse was preferred to his lyrics in literary English.

DUNLAP, William (1766-1839), born in New Jersey, became a portrait painter at 16, and made pastels of George and Martha Washington. While studying in London with Benjamin West he became captivated by the theater, and upon his return he began a career in drama that established him as the first American professional playwright. He also managed theaters, but went bankrupt in 1805. He wrote the *Life of Charles Brockden Brown* (1815). He was a historian of American art, an unsuccessful magazine publisher, an artist, a temperance novelist, and a translator of European drama, as well as the author of 30 plays.

DWIGHT, Timothy (1752-1817), grandson of Jonathan Edwards, one of the "Hartford Wits," a stalwart Calvinist and Federalist, and President of Yale, was the author of sermons, discourses, travel essays, and poems, practically all of which are presently unreadable. However, he was a great leader and teacher.

E

EASTMAN, Max [Forrester] (1883-), the son of two Congregational ministers, was a founder and editor of *The Masses*, an admirer of Trotsky, and a Marxist critic. He was expelled from the Communist Party after his return from a disappointing study of the Soviet Regime in 1922. *Enjoyment of Poetry* (1913) represents his major contribution to literary criticism, having no special reference to Marxian ideology or social study. He is also a historian of Russia and an unsympathetic critic of communism as it has developed in Russia.

EDMONDS, Walter D[umaux] (1903-), graduated from Harvard in 1926, where he had been editor of the *Advocate*. His first novel, *Rome Haul* (1929), a story of the Erie Canal, was later dramatized as *The Farmer Takes a Wife* (1934). He wrote *Drums Along the Mohawk* (1936) in six months, which is typical of his facility in composition, once his historical research is completed.

EDWARDS, Jonathan (1703-58), graduated from Yale in 1720, where he became highly impressed with the work of Locke and Newton. He was appointed pastor of the Northampton church in 1729, where he preached so eloquently on predestination and eternal damnation that some members of his parish attempted suicide. He stimulated a series of revivals and hysterical conversions which reached their peak with the Great Awakening (q.v.) in 1740. Opponents of his extreme Calvinism secured his release in 1750, and he was exiled to Stockbridge, Mass., to minister to the Indians. When his son-in-law, the Rev. Aaron Burr, died (leaving an infant son who was to become famous), Edwards succeeded him as president of the College of New Jersey (Princeton), but died himself two months later of complications from inoculation for smallpox. His many theological publications reveal a paradox of personality and intellect that can only be reconciled, if at all, by considering him as a mystic. He was the founder of "New England theology" and a precursor of Transcendentalism in his negation of the Hebraic God in favor of an infinite being. Among his many works the best known are *Sinners in the Hands of an Angry God* (1741), *Misrepresentations Corrected and Truth Vindicated* (1752), and *Freedom of the Will* (1754).

EGGLESTON, Edward (1837-1902), lived in many Indiana towns during his boyhood as his step-father, a Methodist minister, sought new parishes. Too poor in health to enter DePauw University on a scholarship his own father had provided, he became a Bible agent, a surveyor, a pastor, and a circuit-riding Methodist minister, both in Indiana and Minnesota. In 1866 he became associate editor of the juvenile magazine, *Little Corporal,* and the next year editor of the *National Sunday School Teacher.* He joined the Staff of *Independent* in New York, and in 1871 became editor of *Hearth and Home,* where his first adult novels were published. From 1874 to 1879 he was pastor of the Church of Christian Endeavor in Brooklyn. *The Hoosier Schoolmaster* (1871), his only well known work, presents the dialect and the mores of the early "Hoosiers" with a realism distinguished for its recording of a transitional society. *Roxy* (1878) and *The Circuit Rider* (1874) also satisfy his purpose of making his novels a historical record of civilization in America. Although he believed his later didactic novels and his juvenile biographies to be of greater value than objective history, they are not as acceptable to modern taste as the histories he wrote of the Middle Western culture. His brother, George Cary Eggleston, wrote many romantic novels and books for boys, and was the prototype of *The Hoosier Schoolmaster.*

ELIOT, T[homas] S[tearns] (1888-), being an expatriate American, is currently placed in both the English and American studies of literature. He was born and reared in St. Louis, a city that made little impression upon his poetic imagination, for America is represented by a caricature of Boston, the Eliot family seat. He studied for three degrees and taught philosophy at Harvard, and also attended the Sorbonne and Oxford. Although he completed his dissertation on the idealism of F. H. Bradley, he never bothered to take the Ph. D. degree. In 1914 he established his residence in England, becoming a British subject in 1927. He worked in London as a teacher, bank clerk, and editor—of *The Egoist* (1917-19) and *The Criterion,* which he founded in 1922. He later joined the editorial board of Faber & Faber, Ltd. In 1932 he returned to Harvard for a year as Norton Professor of Poetry. In 1948 he was awarded the Nobel Prize for Literature.

No other poet and critic in the history of English literature has, in his own time, been the subject of more critical controversy and the source of more influence on his contemporary literature than the reserved, publicity-shy Mr. Eliot. He is often called, without qualification, the greatest living poet. His detractors, on the other hand, dismiss him as a contemporary fashion, comparable to Tennyson, or Longfellow. Arguments are conducted in scholarly journals as to whether his criticism contradicts itself, or contradicts his poetry. A whole tradition of criticism evolved to account for him, and to decipher his meanings. The scholarly explanations of his literary references, the painstaking tabulations of his borrowed lines, remind one of research into the work of Dante.

Meanwhile, Mr. Eliot remains, for the most part, aloof to both arguments and explanations, as if he were, like Dante, removed from it all.

Eliot's first poems, printed in the Harvard *Advocate* from 1907 to 1910 (reprinted in Harvard *Advocate* anthology, 1950), reveal his gradual change from the conventional to an approximation of his later style. He was already under the influence of the French *Symbolistes,* particularly of Laforgue and Baudelaire, and in his courses he discovered that Donne, Dante, and Henry James could be useful to him. He heard Professor Irving Babbitt lecture on the superiority of classicism to 19th century romanticism. In London he was welcomed enthusiastically by Pound and Hulme as a fulfillment of the creed they called Imagism. Through Pound's influence, "The Love Song of J. Alfred Prufrock" was published in *Poetry* magazine (1915). His first two volumes of poetry, *Prufrock and Other Observations* (1917) and *Poems* (1919), express in a witty and ironic style his feeling that modern culture is sordid, materialistic, and wholly ignoble. *The Waste Land* (1922), his most erudite, comprehensive, and successful portrait of sterility, draws on more than three dozen sources for quotation or allusion in a mere 433 lines. Using as a framework the Perceval legend, a pagan story of the search for the Christian Grail, he amplifies modern experience by the introduction of visitors and references from the whole gamut of western mythology. *Ash Wednesday* (1930) provides his solution, a return to the "true

Church." His allegiance to Anglo-Catholicism is again expressed in *The Rock* (1934) and in his morality play concerning the murder of Thomas à Becket, *Murder in the Cathedral* (1935). The *Four Quartets* (1943) are mystic, devotional poems carrying even further than *The Waste Land* his theory that "the use of recurrent themes is as natural to poetry as to music." Lacking the wit and clever scepticism of his early work, the religious poetry has not been popular. Although the reader might not always recognize an erudite allusion or follow the narrative in his first volumes, the theme was obvious. When the theme itself becomes abstract and scholarly, many readers lose their patience with obscurity. Through his plays, however, he continues to demonstrate his ability to charm the popular audience: *The Family Reunion* (1939) was successfully performed, even in sterile America, by college groups, and both *The Cocktail Party* (1950) and *The Confidential Clerk* (1954) have been successful on Broadway.

Eliot's essays have been of more value, perhaps, to the understanding of Eliot than of his subjects. They are stimulating manifestoes rather than critical scholarship. *The Sacred Wood* (1920) argues the importance of tradition for literature, while *For Lancelot Andrewes* (1928) and *After Strange Gods* (1934) reveal his increasing conservativism in religion, ethics, and politics. His essays form the basis for the New Criticism. (See CRITICISM, *The New Critics,* for his contributions to modern criticism. See also

"POETRY" for a more general treatment of his influence on contemporary poetry.)

ELLIOTT, Sarah Barnwell (1848-1928), wrote several novels but is remembered for her realistic portrait of mountain life in her native Tennessee, *Jerry* (1891).

ELLISON, Ralph [Waldo] (1914-   ), grew up in Oklahoma, studied at the Tuskegee Institute, and played in dance bands in the Middle West before moving east to pursue his literary interests. After publishing stories in several magazines, he won critical acclaim and the National Book Award for *Invisible Man* (1952), a novel of excellent structure which tells about the experience of being a Negro without bitterness or pathos.

EMERSON, Ralph Waldo (1803-82), was the son of a Unitarian minister and a descendant of Boston "Brahmins", but with the death of the father in 1811, his mother was forced to take in boarders to support the family of six small children. A paternal aunt of eccentric but penetrating mental capacities had considerable influence in formulating the young Emerson's attitudes. He earned his way through Harvard as a messenger and waiter, graduating in 1820. After teaching in his brother's Boston school for young ladies, he studied at the Harvard Divinity School, but interrupted his studies to spend the winter of 1826-7 in Florida because of incipient tuberculosis. In 1829 he became pastor of the Second Church in Boston, and six months later married the beautiful Ellen Louisa

Tucker, who died of tuberculosis within two years. In 1832 Emerson resigned his pastorate because he was troubled by the distinction between religious tenets and everyday life, where he felt no distinction should exist. In 1832 he sailed for a tour of Europe, where he met Carlyle, Coleridge, and Wordsworth who introduced him to Transcendentalism and German idealism. This, in combination with his devotion to Plato and Swedenborg, remained a lasting influence, and the friendship he formed in England with Carlyle endured for forty years of correspondence. On his return to Boston he engaged in sporadic preaching, but turned increasingly to lyceum lecturing on philosophy, culture, and religion. He married Lydia Jackson and settled in Concord, where he became the center of the "Concord School" which included Thoreau, Hawthorne, Bronson Alcott, and Margaret Fuller (qq.v.). Out of his journals, which he kept from boyhood, grew his lectures, and out of both grew the essays he is remembered for today. "Self-Reliance" and "Compensation" are practical recipes for personal integrity. "The Over-Soul" is man's personal intuition of God. "The American Scholar," delivered before the Phi Beta Kappa Society of Harvard, was called by Holmes "our intellectual Declaration of Independence." His "Divinity School Address," which attacked the pretensions of formal theology with merciless logic, alienated Harvard for 30 years. After 1860, when his personal powers were declining, world renown came to Emerson. He was acclaimed a celebrity on his trips to Europe and California, and Harvard awarded him an honorary degree. His dozen or so best poems have a clear, honest style, but in most of his verse he did not practise his own preachings for a break with tradition, and he was embarrassed by the extremes to which Whitman took his recipes for an indigenous American poetry.

ENGLE, Paul (1908-), grew up in Cedar Rapids, Iowa, and attended Coe College. He took his M. A. degree at the University of Iowa, studied for a year at Columbia, and spent three years at Oxford as a Rhodes scholar. His first volume of poems, *American Song* (1934), was acclaimed for its American quality as a literary descendant of Whitman's *Leaves of Grass*. *Break the Heart's Anger* (1936) followed the fashion of denouncing American materialism, but he returned to the idiom of Iowa farm life with *Corn* (1939). He also wrote the novel, *Always the Land* (1941). He is resident poet and a teacher at the University of Iowa.

ENGLISH, Thomas Dunn (1819-1902), studied both law and medicine, but became a journalist, and wrote several novels and plays. He is remembered as the author of the famous ballad, "Ben Bolt," which Du Maurier used in *Trilby*, and has since been set to music 24 times. He was an intimate of Poe, but the two became enemies, and Poe caricatured him as "Thomas Dunn Brown" in "The Literati."

ERSKINE, John (1879-1951), was reared in New York City and took his B.A., M.A., and Ph.D. degrees at Columbia, where he later became a professor of English. In addition to functioning as teacher, musician, scholar, poet, and critic, he wrote several sophisticated, humorous versions of famous legends: *The Private Life of Helen of Troy* (1925), *Galahad* (1926), *Adam and Eve* (1927), and *The Brief Hour of François Villon* (1937). He was co-editor of the *Cambridge History of American Literature,* and his doctoral dissertation, *The Elizabethan Lyric,* has received four printings. As a pianist he was soloist with the New York, Detroit, Minneapolis, and Chicago Symphony Orchestras.

EVANS, Charles (1850 - 1935), Boston librarian, edited *American Bibliography* (12 vols., 1903-34), a catalogue of the books published in America from 1639 to 1799, and containing 35, 854 titles.

EVANS, Nathaniel (1742-1767), was born and reared in Philadelphia, the son of a merchant. After study at the Philadelphia Academy he traveled to England, where he was ordained a clergyman. Before his death from tuberculosis at 25, he wrote poems, imitative of Milton and Gray, which revealed a remarkable gift for grace and melody.

Expressionism, a critical term, used in painting to designate such extreme departures from reality as *cubism, futurism,* etc., is also applied to that literature which presents improbable or impossible action within a spirit of reality, as opposed to fantasy. In his realistic play, *The Hairy Ape,* Eugene O'Neill has the tough seaman designated by his title go up to some elegant gentlemen to shoulder them off the walk, but it is the underprivileged sailor who bounces back. Thereby O'Neill suggests that in any less physical encounter between the two social classes, the poor and the uneducated will be shoved aside. The Expressionism of Joyce or Eliot, however, cannot be defined in these simple terms, and is characterized by the domination of the artist's mind over his subject matter.

F

FARRELL, James T[homas] (1904-   ), excelled in athletics while attending Catholic schools in Chicago. While studying at the University of Chicago, he wrote a short story, "Studs," which he later developed into the *Studs Lonigan* trilogy (1935), a Naturalistic, almost sociological study of a young Catholic on Chicago's South Side. His adolescence is portrayed in *Young Lonigan* (1932); *The Young Manhood of Studs Lonigan* (1934) takes him into the underworld; and he meets his death in *Judgment Day* (1935). With *A World I Never Made* (1936), Farrell began the story of Danny O'Neill, more sensitive than Studs, continued in *No Star Is Lost* (1938), *Father and Son* (1940), and *My Days of Anger* (1943). In addition to these and several other novels, he has published seven volumes of short stories.

A stern moralist, Farrell writes consistently with the purpose of revealing what he sees as the immorality in contemporary life, by giving examples in fiction rather than moralizing or propagandizing within fiction. His political philosophy is a curious blend of Pragmatism and Marxism. He bitterly opposes Stalinist Marxism, and is bitterly opposed, in turn, by the Russian Communists. He believes that man should be free in the fullest sense of that word, and that man loses his freedom through capitalistic institutionalism, his individual force being weaker than the forces of the institutions. He recognizes that the Communism practised in the U.S.S.R. also deprives the individual of freedom, but he does not recognize that this is inevitable when Marxism is brought to practical application. His political attitudes are most clearly presented, although they are not clearly reconciled, in *Reflections at Fifty* (1954).

FAST, Howard [Melvin] (1914-   ), has published numerous historical novels, of which the best known is *Citizen Tom Paine* (1943).

FAULKNER, William [Harrison] (1897-), is the only American novelist who makes as many demands of the reader as do our modern poets. Never a popular author, he worked in isolation from literary people or movements for years, steadily evolving his own saga of the South in his own way, praised by only a few critics who perceived his plan, and more often condemned for his "technical virtuosity" or for his inclusion of incidents that violate even the mores of this century. Gradually, more champions appeared to shock conservative critics with the announcement that Faulkner is our greatest American novelist, a movement of critical opinion which reached a crescendo when he received the *Nobel Prize* for literature in 1950.

Faulkner grew up in Oxford, Mississippi, the setting of most of his work, and his home until recent years. He served with the British Royal Air Force in World War I, and was wounded in a plane accident. During his spare time he attended courses at Oxford University. Never a successful student, he took only a few courses upon his return, at the University of Mississippi where his father was business manager. After the private publication of *The Marble Faun* (1924), a book of pastoral poems, he drifted to New Orleans, and shared an apartment with Sherwood Anderson, who became a major influence on his prose.

In 1926 he published *Soldier's Pay,* a novel about a dying aviator returning to a shallow southern belle, and marrying instead a war widow who fulfills his needs. *Mosquitoes* (1927) places a group of New Orleans aesthetes on a disabled yacht in a swamp, and records their Bohemian conversation. Returning to Oxford, Miss., he wrote *Sartoris* (1929), which chronicles the deterioration of the older aristocratic families with the rise of such upstarts as the Snopes clan. *The Sound and the Fury* (1929) portrays the degeneration of the Compton family, partly through the stream of consciousness of their idiot son.

Degeneration became the theme of many of the later novels, in which he sometimes developed more fully incidents or characters depicted in *Sartoris.* This was, in fact, the theme of the Faulkner (or Falkner) family, whose ancestors had been statesmen, giants of southern industrial expansion, or generals. William's great grandfather, Colonel William Falkner, was the author of *The White Rose of Memphis,* which sold 160,000 copies. William might well envy his grandfather, for even after the publication of six books the grandson had to support himself by shoveling coal in a power plant on the night shift while, in his free time, he wrote *As I Lay Dying* (1930) on an upturned wheelbarrow. One of his most compact novels, this tragicomedy portrays a mountaineer family enduring first the death of the mother, and then a frustrating search in floodtime for a passage to the burial ground in town, hauling the corpse on a wagon. The Bundrens are not wilfully perverse; they are simply subnormal and incredibly irrational. With *Sanctuary* (1931), Faulkner attempted to produce a degeneracy so horrifying that the public would be shocked into buying his book. It was his only novel, until the printing of pocketbook editions, to sell more than 10,000 copies. *Light in August* (1932) is constructed in what Richard Chase has called "a poetry of physics." A pregnant girl, in search of her lover, personifies light, pursuit, the life force, and the curve of tradition. The linear Joe Christmas is in flight to break out of the circle, and finds death. It is typical of

Faulkner to avoid so obvious a construction as to cast these two as the lovers; they never meet, but the girl's lover is Joe's weak friend. *Absalom, Absalom!* (1936) leaves only a half-breed moron as a memorial to a mountaineer's dream of establishing an aristocratic family. Faulkner's other books include *Pylon* (1935), showing barnstorming pilots at a Mardi Gras; *The Unvanquished* (1938), stories of the Sartoris family; *Go Down Moses* (1942), short stories; *Salmagundi* (1932), essays and poems; *The Wild Palms* (1939) presenting a convict, a doctor, and his mistress; *The Hamlet* (1940), depicting the rise of the Snopes family; *Intruder in the Dust* (1948), concerning the Southern problem of racial relations; and a religious allegory concerning war, *A Fable* (1954). Recently Faulkner has worked for the motion picture industry in Hollywood as a scenarist, and with television in New York where some of his stories were produced.

FAY, Theodore Sedgwick (1807-98), diplomat, poet and essayist, wrote articles for the New York *Mirror* which were published as *Dreams and Reveries of a Quiet Man* (2 vols., 1832). His novel, *Norman Leslie: A Tale of the Present Times* (1835) was severely criticised by Poe.

FEARING, Kenneth (1902-   ), was born in Oak Park, Ill., and after graduation from the University of Wisconsin in 1924, entered journalism in Chicago. Shortly thereafter he moved to New York, where he has done free-lance writing and engaged in editorial and publicity work. His six volumes of poetry contain satires of middle class mores and attitudes, written in conversational idiom with considerable humour: *Angel Arms* (1929), *Poems* (1935), *Dead Reckoning* (1938), *Collected Poems* (1940), *Afternoon of a Pawnbroker* (1943) and *Stranger at Coney Island* (1949). His first novel, *The Hospital* (1939), is a New York kaleidoscope through an X-ray lense. *Dagger of the Mind* (1941) is a murder mystery, while *The Big Clock* (1946) portrays a man at odds with society in search for himself.

Federal Theatre Project (1936-9): a W. P. A. relief program for the unemployed in the profesional theater. Units were established throughout the country to produce experimental plays in the little-theater tradition, usually for audiences who had never seen a stage play. One of the innovations was termed, "The Living Newspaper," a dramatized presentation of social problems of the day.

Federal Writers' Project (1935-9): a W. P. A. program for the relief of unemployed journalists, editors, and research workers throughout the United States. The major project was a compilation of geographical and cultural information about each specific area, called the "American Guide Series."

FENNELL, James (1766 - 1816), was born in London and attended Eton and Cambridge before coming to America, where he became known as an actor and playwright. The best of his plays was the comedy, *Lindor and Clara: or, The British Officer* (1791).

FENOLLOSA, Ernest Francisco (1853-1908), a pioneer in the study of Oriental art, history, and literature, was born in Salem, Massachusetts, studied philosophy at Harvard, and in 1878 became a professor at the University of Tokio, where he remained for twelve years. On his recommendation, purely Japanese art with the use of ink, brush, and paper was reintroduced into the schools of Japan, and he also suggested that they cease their imitation of European art. He wrote *East and West: The Discovery of America and Other Poems* (1893) and *Epochs of Chinese and Japanese Art* (1912, compiled by his wife). His literary executor, Ezra Pound, wrote the poetic translations, *Cathay* (1915), from Fenollosa's notes, and edited *Certain Noble Plays of Japan* (1916), and *Noh—or, Accomplishment* (1916).

FERBER, Edna (1887-), grew up in the Middle-West and, after graduation from high school, became a reporter in Appleton and Milwaukee, Wisconsin. She first attracted attention with her short stories about Emma McChesney, a traveling petticoat saleswoman. Of her many novels, *So Big* (1924) won a Pulitzer Prize, and *Show Boat* (1926) has been produced on stage, screen, and radio. With George Kaufman she wrote the dramatization of *Show Boat, The Royal Family* (1927) and *Dinner at Eight* (1932). Columbia University awarded her an honorary doctorate of letters.

FERGUSSON, Harvey (1890-) a New Mexico novelist, worked for the Forest Service after graduation from Washington and Lee University and later became a reporter in Washington, Savannah, and Richmond. His trilogy, *Followers of the Sun*, presents historical fiction set in the Southwest, particularly in New Mexico: *Blood of the Conquerors* (1921); *Wolf Song* (1927), produced as one of the first realistic western motion pictures; and *In Those Days* (1929).

FERRIL, Thomas Hornsby (1896-), is the first major poet to use western history and the Rocky Mountain setting without succumbing to the common fallacy, predicted for this region by Oscar Wilde in 1882, of sentimentalizing the grandeur of inanimate scenery, a conceit that leaves man dwarfed by mountains. Ferril uses scientific awareness to keep the mountains in their place. Born in Denver, the fifth generation of pioneers, he attended a local Latin school and graduated from Colorado College. During the First World War he was commissioned a second lieutenant in the Air Service after training at Columbia University. Following a career as drama critic for the Denver *Times*, he became publicity director for the Great Western Sugar Company, editing company publications and producing industrial and agricultural motion pictures. His first book, *High Passage* (1926), won the Yale Series of Younger Poets competition. The following year he was awarded *The Nation's* poetry prize for "This Foreman," and in 1936 *Poetry* Magazine awarded him the Oscar Blumenthal prize for "Words for Leadville." In *Westering* (1934), he broadened his canvas, writing longer poems to cover a longer dimension of time and space while experimenting with symphonic rhythms to match his subjects. *Trial by Time* (1944) reveals an extension of his experimentation. The poems show a "tighter" manipulation of both prosody and intellectual observation, and range in length from the 229 lines of "Words for Leadville," which chronicles the geological as well as the anthropological history of a mining town, to one of the greatest condensations of meaning in poetry: "Planet skin,/ Is festering pink,/ Where protoplasm,/ Learned to think." This tendency toward classic economy and restraint is carried even further in his later work published in *New and Selected Poems* (1952), which won the Ridgely Torrence Memorial Prize. Although he uses a western frame of reference, his theme is time, stretching back for geological eons, or stretching forward to "A child far-generated, lover to lover . . . lover to lover over." He fits into no contemporary critical category, in form or content. His prosody varies from the tight lyric to a casual vernacular according to the demands of the subject. His symbolic references range from Attic to western to scientific. The verses which could submit to pages of explication are still intelligible on their primary level. Neither an optimist nor a pessimist, he observes the strange coincidences of human experience with the passionate wonder of a scientist. Having achieved his potential strength long after he was first recognized, and having re-

jected contemporary fashions in his progression, Ferril was somewhat neglected by eastern critics after their initial enthusiasm for his early work, but the younger critics are now beginning to take an interest in his later work. He has received honorary degrees from the University of Colorado, the University of Denver, and Colorado College. Cecil Effinger's *Symphony for Chorus and Orchestra*, based on Ferril's poem, "Words for Time," was recently performed in New York and Colorado. Bernard DeVoto, who regards him as "the best poet now in active practice," also praises him as "one of the best writers of prose anywhere." Ferril's informal essays, usually attacking "sacred cows," have been published in his family paper, *The Rocky Mountain Herald*, and in *The Saturday Review of Literature*, the New York *Herald Tribune*, and *Harper's Magazine*, for which he wrote the bi-monthly column, "Western Half Acre." His essays were collected as *I Hate Thursday* (1946), with illustrations by his daughter, Anne Ferril Folsom. (See also, POETRY).

FESSENDEN, Thomas Green (1771-1837), New Hampshire lawyer, inventor, and editor, wrote scurrilous attacks on Jefferson and other leading democrats in bitter verse, collected as *Democracy Unveiled* (1805). His *Original Poems* (1804) included "The Country Lovers," sometimes suggested as Lowell's model for "The Courtin'." Fessenden wrote under the pseudonym, Christopher Caustic.

FICKE, Arthur Davison (1883-), an

Iowa poet whose early volumes of romantic poetry include *From the Isles* (1907), *Sonnets of a Portrait Painter* (1914) and *The Man on the Hilltop* (1915). With Witter Bynner (q.v.) he wrote the satire on modern poetry, *Spectra* (1916).

FIELD, Eugene (1850-95), a St. Louis journalist who conducted the "Sharps and Flats" column in the Chicago *Daily News* from 1883 to 1895, containing humorous verse, often in imaginary dialect. While writing, earlier in his career, for the Denver *Tribune*, he was known for his practical jokes. When Oscar Wilde visited Denver, Field disguised himself as the flamboyant Englishman, stepped off an earlier train, and graciously acknowledged the enthusiasm of people lining the streets as he rode in a carriage from the station. He is best remembered for his juvenile verse, "Little Boy Blue," and "Wynken, Blynken, and Nod" ("Dutch Lullaby"), but he published several volumes of the material from his column, including *A Little Book of Western Verse* (1889) and *The Love Affairs of a Bibliomaniac* (1896).

FIELD, Rachel [Lyman] (1894-1942), Massachusetts dramatist and novelist, wrote *The Cross-Stitch Heart and Other One-Act Plays* (1927) and many other plays for children. Her novels (for adults) are *Time Out of Mind* (1935), *All This, and Heaven Too* (1938), and *And Now Tomorrow* (1942).

FIELDS, Annie [Adams] (1834 - 1915), married James T. Fields (q.v.) at twenty, and became famous for her Charles Street *salon* in Boston, and for her gay and intellectual personality as its hostess. In addition to several volumes of verse, she wrote amplifications of her journals, including *Authors and Friends* (1896). M. A. DeWolfe Howe edited a selection from her journals in 1922: *Memories of a Hostess*. After her husband's death Mrs. Fields toured Europe with Sarah Orne Jewett (q.v.)

FIELDS, James T. (1817-81), edited the *Atlantic Monthly* (1861-71) and wrote several volumes of poetry, including *A Few Verses for a Few Friends* (1858). His book of essays, *Yesterdays with Authors* (1872) is considered more rewarding today than the poems, for it recalls Fields' many associations with the prominent authors of his time. As the major figure in the famous publishing house of Ticknor and Fields he published most of the important British and American authors of the day. He was a member of the famous Saturday Club in company with Emerson, Hawthorne, Holmes, Lowell and Longfellow, and his wife conducted the best known literary *salon* in Boston.

FINK, Mike (1770?-1823?), a keelboatman on the Ohio and Mississippi Rivers, whose adventures were elaborated in the oral tradition, after his death, to give him a fabulous stature as one of the American myths.

FINLEY, Martha (Martha Farquharson) (1828-1909), wrote the 28 Elsie Dinsmore books (1867-1905) in addition to many other juvenile novels, incredibly sentimental propaganda pieces to inculcate smugness and piety in the innocent child reader.

FISHER, Dorothy Canfield: See Canfield.

FISHER, Vardis [Alvero] (1895-), was born and reared in Idaho, took his B.A. degree at the University of Utah, and his M.A. and Ph. D. degrees at the University of Chicago. Most of his early experiences are delineated in the autobiographical tetralogy: *In Tragic Life* (1932), *Passions Spin the Plot* (1934), *We Are Betrayed* (1935), and *No Villain Need Be* (1936). In 1939 he received the Harper Novel Prize for *Children of God,* a historical novel of the Mormons. His other works include historical novels concerning the pioneering of the West, and a series concerning prehistoric man. His poems were published as *Sonnets to an Imaginary Madonna* (1927), and his essays as *The Neurotic Nightingale* (1935). He has taught English at the University of Utah and at New York University, and directed the Federal Writers' Project in Idaho.

FITCH, Clyde [William] (1865-1909), grew up in Schenectady, and at Amherst was known for his clever writing and his many pranks. In 1888 he went to Paris, where his interest in the theater received a permanent stimulus. On his return he received the fortunate assignment to write a play for Richard Mansfield, and produced *Beau Brummel,* an immediate success. Of his 30 plays, many others were also written for a specific star.

Barbara Frietchie (1899) was created for Julia Marlowe. He seemed to have equal facility with farce, history, social drama, and contemporary problem plays. His great technical facility and his personal showmanship apparently aroused the hostility of the critics, who frequently misunderstood his more serious intentions. *The Truth* (1906) won much higher praise in Europe than in New York. After a long period of poor health, he died while journeying through the Tyrol in 1909.

FITZGERALD, F[rancis] **Scott** [Key] (1896-1940), was born in St. Paul, Minnesota, attended preparatory school in New Jersey, and at Princeton wrote poems and stories for the college magazines, as well as part of the libretto for a Triangle Club production. Edmund Wilson, John Peale Bishop, and Charles Bayly were among his fellow students. He served in World War I as a lieutenant, and wrote *This Side of Paradise* in three months of his off duty time. Failing to find a publisher for it after the war, he rewrote the novel in two months, and it was published in 1920. A cynical portrait of the jazz age, it established his reputation for sophisticated wit, and he was able to earn an excellent living thereafter by writing stories for the popular magazines, collected as *Flappers and Philosophers* (1920) and *Tales of the Jazz Age* (1922). His second novel, *The Beautiful and Damned,* was published in 1922, and a satirical play, *The Vegetable; or, From President to Postman,* in 1923. Fitzgerald wrote well only when he worked exceedingly hard through countless revisions. So far he had had nothing to say that he felt was worth the trouble. He wrote six drafts of his next novel, and for the first time developed his latent talent for dramatic narrative. *The Great Gatsby* (1925) has one of the most perfect structures in literature. It has the Fitzgerald trademark, an atmosphere of glamour and romance, a dialogue of understatement reinforced by rich, concrete, sensuous imagery, and a cast of incredibly wealthy characters. The narrator, an observer from merely the upper-middle class, allows Fitzgerald to objectify the play of emotions. The protagonist, Major Jay Gatsby, is discovered by an unscrupulous operator, right after the war, sleeping on pool tables in his uniform. As the novel opens he is living on a huge estate at West Egg, Long Island, where he entertains lavishly. He hopes to entertain the girl he fell in love with during the war, now married to a brutal boor of great wealth and residing across the bay. Fitzgerald describes Gatsby, and the book, in his opening pages: "—Gatsby, who represented everything for which I have an unaffected scorn. If personality is an unbroken series of successful gestures, then there was something gorgeous about him, some heightened sensitivity to the promises of life . . . it was an extraordinary gift for hope, a romantic readiness . . . No—Gatsby turned out all right at the end; it was what played on Gatsby, what foul dust floated in the wake of his dreams that temporarily closed out my interest in the abortive sorrows and short-winded elations of

men." Perhaps these lines also describe Fitzgerald, to some degree. During the twenties he and his wife, Zelda, exhibited an impatient readiness to live the books he portrayed. His later books indicate that he too met the ultimate dissolution. *Tender Is the Night* (1934) was to be his masterpiece, but it was not appreciated until after his death. *The Last Tycoon* (1941), an unfinished novel, concerns a motion picture executive in Hollywood, where Fitzgerald worked the last few years of his life. *The Crack-Up* (1945) contains essays, notes, and letters in addition to critical comments by other authors. During the depression he was remembered, if at all, merely as the man who had written about the tennis set. Now a new interest has resulted in biographies and critical studies that promise to establish his position as one of our best novelists, at his best.

FLETCHER, Inglis (1888-   ), is noted as the author of the monumental 7-volume Carolina Series of novels of the Colonial South, beginning with *Raleigh's Eden* (1940) on the causes of the Revolution in North Carolina. *Men of Albemarle* (1942), on the evolution of law and order; *Lusty Wind for Carolina* (1944), on trade and piracy; *Toil of the Brave* (1946), on the critical contests of the Revolution; *Roanoke Hundred* (1948), on Sir Walter Raleigh's first unsuccessful colony; *Bennett's Welcome* (1950), on the first permanent settlements in North Carolina; and *Queen's Gift* (1952), on the ratification of the Constitution—fill out the Series.

FLETCHER, John Gould (1886-1950), Arkansas poet, was educated at Harvard, but left in his senior year, upon receipt of a substantial inheritance from his father, to write and study informally in Boston, London, and Paris. In London he was associated with the Imagists and published five volumes of poetry in 1913. *Irradiations* (1915) and *Goblins and Pagodas* (1916) were published in America. He lived in England from 1916 to 1933, when he returned to Arkansas and became a leader among the Southern Agrarians (q.v.). His *Selected Poems* (1938) were awarded the Pulitzer Prize. He has also written a biography of *Paul Gauguin* (1921), translated Rousseau, and published critical pamphlets. *Life Is My Song* (1937) is his autobiography.

FLINT, Timothy (1780-1840), a Massachusetts missionary, wrote several reports of the western frontiers he had visited, romanticizing his histories in his enthusiasm for explaining the West to the East. His novels and histories include: *Francis Berrian; or, The Mexican Patriot* (1826), *George Mason, the Young Backwoodsman* (1829), *The Shoshonee Valley* (1830), *Indian Wars of the West* (1833) and *The Biographical Memoir of Daniel Boone* (1834).

FOERSTER, Norman (1887-   ), the son of a Pittsburgh composer, was a pupil of Willa Cather in high school, and of Irving Babbitt at Harvard, whom he soon joined in promoting the new Humanism (q.v.). He has taught at the Universities of Wisconsin, North Carolina, and Iowa. He is the author of critical essays and books: *Nature in American Literature* (1923), *American Criticism* (1928), and *Toward Standards* (1930).

FORD, Paul Leicester (1865-1902), novelist, bibliographer, and historian, grew up in Brooklyn, N. Y. The fact that he was a hunchback and in delicate health prevented his attending public school, and he received his education from private tutors. His father, a book collector, had a magnificent library containing rare historical American materials, which Paul began to study at an early age. With additional research in Europe, he was superbly qualified for his historical scholarship, which included *The Writings of Thomas Jefferson* (10 vols., 1892-4), and *The True George Washington* (1896). Of his many popular novels, the best remembered are *The Honorable Peter Stirling* (1894) concerning petty politics, and *Janice Meredith* (1899), which sold 200,000 copies in three months. At 37, he was shot by a disinherited brother.

FOSTER, Hannah Webster (1759-1840), the wife of a Massachusetts minister, wrote one of the earliest American novels, *The Coquette* (1797), based on an actual scandal of seduction. She named all of the participants except the victim, which may account in part for the book's extreme popularity.

FOX, John [William], Jr. (1863-1919), grew up among the Cumberland mountaineers of Kentucky, who became the main subjects for his stereotyped novels.

Although less sentimental than his early work, *The Little Shepherd of Kingdom Come* (1903) and *The Trail of the Lonesome Pine* (1908) achieved an enormous popularity.

FRANK, Waldo [David] (1889-   ), was born in New Jersey, the son of a New York lawyer, and began writing before the age of 5. A novel was accepted by a publisher when he was 16, but his father withdrew it. After a year of school in Switzerland, he attended Yale, ranched in Wyoming, worked for newspapers in New York, and in 1914 settled down to writing as a profession. He has traveled widely, and is an authority on South American culture. Often identified with radical movements, he believes in his own definition of socialism, but rejects both Marxism and Stalinism, which he attacked in *Chart for Rough Water* (1940). His political philosophy and his novels include an aesthetic vision of mystical values, and he favors the poetic novel over realism. But unlike most poetic novelists, he immerses the individual in his time, place, and society. His major novels are *The Death and Birth of David Markand* (1934), concerning a business man embracing Marxism; *The Bridegroom Cometh* (1939), which includes a woman's disenchantment with communism; *Island in the Atlantic* (1946) covering three generations of New York City history; and *The Invaders* (1948). He has also written several interpretations of American and Spanish culture.

FRANKLIN, Benjamin (1706-90) is difficult to place in our literary tradition. One of our most important writers, he wrote some of our least important creative work, while turning out volumes of important scientific scholarship, political commentary, and practical observation on almost every phase of human behavior. Born in Boston, he was apprenticed to his father at the age of ten, but so disliked the tallow-making trade that at twelve he was sent to his half-brother, James, printer and later publisher of the *New England Courant*, in which Franklin's "Dogood Papers" first appeared. Although he received only two years of elementary education, he read widely in Xenophon, Locke, Shaftesbury, and Addison. Because of constant friction with James, he ran off and engaged in printing in Philadelphia. Governor Keith proposed to set him up in business, and in 1724 sent him to London to buy equipment, but failed to send the expected money. Franklin worked as a printer in London for two years to earn passage back to Philadelphia. In 1728 he bought the *Pennsylvania Gazette*, which he transformed into an interesting and successful publication, partly by his own contributions. They included the *Busybody Papers*, "Dialogues between Philocles and Horatio," and *Poor Richard's Almanack* (1732-58), which contained aphorisms borrowed from Bacon, Rabelais, and Rochefoucauld. In 1730 he married Deborah Read, bringing to his new home his illegitimate son, William. As his business enterprises prospered, he became interested in the affairs of the city. He founded the Philadelphia fire company, the Pennsylvania Hospital, the Junto club, the Academy which later became the University of Pennsylvania, and the American Philosophical Society. He initiated paving, street-cleaning and lighting, a city police, and the first circulating library. He invented the Franklin Stove and a new type of clock, and conducted experiments in electricity. By 1750 he had begun his political activities, and in 1757 was sent to England to negotiate tax concessions. He remained there until 1774 as agent, virtually as ambassador, for Pennsylvania, New Jersey, Georgia, and Massachusetts. After his return, he helped frame the Declaration of Independence. In 1776 he was sent to France to secure aid for the American Revolution, and remained as plenipotentiary until 1785, becoming one of the most honored and loved men in all France. As a writer, Franklin was most at his ease in the essay. At times he used irony in behalf of political convictions ("Edict by the King of Prussia," 1773; "Rules by which a Great Empire may be reduced to a Small one," 1773). Or, he could be practical, witty, urbane, and familiar in conversation pieces such as "The Dialogue between Franklin and the Gout" (1780). His unfinished *Autobiography* (1818) reveals his clarity and precision as a writer, flavored with wit and genial cynicism. Too busy ever to finish a full-length book, he is America's "universal man," though not a representative of "universal America." He lacked the idealism of the New England mystics, the poetry of Whitman or Melville. He stands as the personification and perhaps as the father of American pragmatic,

middle-class, self interest in its greatest devotion to the welfare of the community as a whole.

FREDERIC, Harold (1856 - 98), New York novelist and journalist, was the son of a railroad worker killed in a wreck when the boy was two. He helped his mother run a dairy until, at 14, he left school to work at office jobs in Utica. He worked in Boston as a photographic retoucher, but because of eye trouble returned to Utica, and at 19 became a reporter on the *Utica Observer*. At 28 he became London correspondent for *The New York Times,* and remained in that position for the remainder of his life. His first novel, *Seth's Brother's Wife* (1887), is a local-color portrait of the life he knew in the Mohawk Valley. *The Copperhead* (1893) depicts the intolerance of New York Abolitionists toward a local farmer who did not share their views. His best and most popular novel, *The Damnation of Theron Ware* (1896), is a realistic chronicle of a Methodist minister who comes to have doubts concerning his doctrines while hungering for the more intellectual society of the town, which he is unfit to enter. Frederic was a careless author and did not revise his first drafts, but he had considerable talent for accurate detail and the energetic development of both his characters and situations.

FREEMAN, Mary E[leanor] Wilkins (1852-1930), Massachusetts writer of the local-color school, wrote historical and social novels, a play concerning the Salem witchcraft trials (*Giles Corey,*

F. SCOTT FITZGERALD

ROBERT FROST

Photo: Lee Samuels

ERNEST HEMINGWAY

HENRY JAMES

Yeoman, 1893), and tales about rural New England life during a period of decay. *A New England Nun* (1891) contains her best stories.

FRENEAU, Philip [Morin] (1752-1832), the son of a wealthy New York wine importer, was educated by private tutors before entering Princeton at 15, where he was a classmate of Madison and Brackenridge. He spent two years as secretary to a planter in Santa Cruz, Danish West Indies, where he wrote his best lyrics, including "The Beauties of Santa Cruz" and "The Jamaica Funeral." Although he never fought in the Revolution, he was twice captured by the British at sea, once on his return from Santa Cruz, and again in 1780 as he attempted to ship again to the Caribbean. After working in the Philadelphia post office, he sailed his own brig from 1784 to 1789. He then engaged in journalism in support of Jefferson until 1795, when he returned to the sea. In 1807 he retired to his New Jersey estate, "Mt. Pleasant." His humorous and satirical political verse won him a reputation during his own lifetime, but his romantic lyrics of the sea, nature, or the Indians establish his uneven genius.

FROST, Robert [Lee] (1875-), was born in San Francisco, but when he was ten his father died leaving his mother without funds, and she moved back to her family home in Lawrence, Massachusetts. He sold his first poem at 14, and was valedictorian of his high school class. He left Dartmouth after a few months to become a bobbin boy in a mill, a Latin teacher in his mother's school, and an editor of a weekly paper. He married at 20, and at 22 entered Harvard, but after two years of specialization in the classics, he found the academic discipline too irksome, as he had found it at Dartmouth. He went back to teaching and editing, and even tried shoemaking. In 1900 he moved to a New Hampshire farm. He taught English at the Pinkerton Academy (1905-11) and psychology at the New Hampshire State Normal School (1911 - 12). Meanwhile he constantly wrote poetry in his simple, honest, conversational style, but publishers continued to reject his submissions, sometimes even inviting him to submit no more. Feeling that his work might have a better reception in England, he sold his farm and moved there in 1912. Living in the West Country, he became the friend of Lascelles Abercrombie and Wilfrid Gibson, and even Ezra Pound became interested in his work. The two books published there, *A Boy's Will* (1913) and *North of Boston* (1914), received enthusiastic reviews. When he returned in 1915 to settle on another New Hampshire farm, he found himself famous, and the publishers who had long rejected his poems now begged for them. In 1916 he became professor of English at Amherst, where he taught until 1938, with interruptions to be honored as poet in residence at Michigan, or to spend winters in Coral Gables, Florida where he lectured at the University of Miami. He has received more honorary degrees from leading universities than a President. He has been awarded the Pulitzer Prize for *New Hampshire* (1934), *Collected Poems* (1930), and *A Further Range* (1936). In 1916 he was Phi Beta Kappa poet at Harvard, and in 1939 he received the gold medal of the National Institute of Arts and Letters. Throughout all this adulation, Frost appears to enjoy the irony of his sudden fame more than the glory. He continues to write as he pleases, sometimes mocking the new critical attitudes with gentle satire. For him, poetry is essentially dramatic. It is a performance, like an athletic endeavor. He is a metaphysical poet who never raises his voice or words to elocution. What appears on casual reading to be merely a verse narrative of an incident or anecdote gradually surrounds itself with a mystery of implication, and the poetry exists in the tension between the expressed and the unexpressed or inexpressible. As with his *West-Running Brook* (1928), he likes to run contrary to popular opinions or attitudes, to take "the road less traveled by." His technique is deliberately conceived to realize "the sentence sounds that underlie the words," and though he writes in verse and rhyme, he has come further than Whitman in realizing Emerson's "Monadnock" call for an unaffected language. One of the new poets, Frost will never fit into any of the categories of new poetry, and might be called a natural bridge between old and new attitudes, merely touching each shore. (See also, POETRY).

Fruitlands: a collective community established at Harvard, Massachusetts (1842-3) by Bronson Alcott, and later described by his daughter, Louisa, in *Transcendental Wild Oats.*

Fugitive, The (1922-5), a bimonthly "little" magazine which was published by several faculty members of Vanderbilt University in Nashville, Tennessee, including J. C. Ransom, Allen Tate, Laura Riding, Donald Davidson, and R. P. Warren. It was essentially a literary review and journal, containing poetry and criticism, but it also was critical of aristocratic aspects of the older South as they emerged to influence the contemporary scene.

Fugitive Group: a term used to identify a group of Southern Agrarians who published *The Fugitive* magazine (q.v.) in Nashville, Tennessee.

FULLER, Henry Blake (1857-1929), a Chicago novelist who wrote early realistic novels, the most successful being *The Cliff-Dwellers* (1893), concerning problems of social and economic caste in the banking business. *With the Procesession* (1895) also treats the striving for social equality. *The Last Refuge* (1900) represents his completely different type of novel, the romantic fantasy of courtly Europe. Others in this vein are *The Chevalier of Pensieri-Vani* (1890) and *The Châtelaine of La Trinité* (1892).

FULLER, [Sarah] Margaret (1810-50), the Massachusetts feminist, was accepted by the Transcendentalists (q. v.) as an intellectual equal. Educated by her father, she read Latin at 6 and Greek as well as French and Italian at 15. She taught school and conducted educational "conversations" for Boston society women. Her book, *Woman in the Nineteenth Century* (1845), considered feminism in its economic, intellectual, political and even sexual aspects. She edited *The Dial* for two years, and later wrote for the New York *Tribune,* including letters from abroad. In England she met her literary idols, Wordsworth, Sand, Carlyle, and the Brownings. In Italy her friendship with the liberator, Mazzini, led to a love affair with Giovanni Angello, Mar-

quis Ossoli, whom she probably married in 1849. After helping fight the cause of liberation, they set sail for America in May, 1850, but the ship was wrecked on Fire Island, and only the body of their child was found. The manuscript of her completed history of the Roman Revolution was lost. *At Home and Abroad* (1856) and *Life Without and Life Within* (1859) were published posthumously. She is considered to be the prototype of Zenobia in Hawthorne's *The Blithedale Romance,* of Miranda in Lowell's *Fable for Critics,* and of *Elsie Venner* by Holmes.

G

Galaxy, The (1866-78), a monthly literary magazine which contained historical, political, and scientific articles as well as fiction and literary criticism. Henry James, J. W. DeForest, Whitman, E. C. Stedman, and Mark Twain were among the contributors. Founded as an antidote in American letters to the "provincialism" of the *Atlantic Monthly,* it was forced to sell its subscription list to the *Atlantic* when it failed to achieve financial success.

GALE, Zona (1874-1938) grew up in Portage, Wisconsin, and attended the University of Wisconsin, taking both the B. A. and M. A. degrees. After newspaper work in Milwaukee and New York, she returned to Portage, determined to achieve success as a writer. Her local-color stories soon found publishers, and were collected as *Friendship Village* (1908), *Yellow Gentians and Blue* (1927), and *Bridal Pond* (1930). Her dramatization of her novel, *Miss Lulu Bett* (1920), was awarded the Pulitzer Prize in 1921.

GARLAND, [Hannibal] Hamlin (1860-1940), grew up on the Middle-Western farms that became the subject of his stories, realistic in their detail of labor and odors, but highly romantic in theme. They were collected in *Main-Travelled Roads* (1891), *Prairie Folks* (1893), *Wayside Courtships* (1897), and *Other Main-Travelled Roads* (1910). While living in Chicago he founded the Cliff Dwellers (1907), a literary organization, and was intimate with Henry B. Fuller, Eugene Field, and Lorado Taft—who became his brother-in-law. In 1915 he moved to New York, believing, correctly, that there he would find greater recognition. His autobiography was published as *A Son of the Middle Border* (1917), and *A Daughter of the Middle Border* (1921), which was awarded the Pulitzer Prize. During 1892 he published three propaganda novels: *Jason Edwards: An Average Man* for Henry George's Single Tax theory, *A Spoil of Office* for the Populist party, and *A Member of the Third House* against the railroad lobby. In his last years he wrote several books of memoirs. A great friend and admirer of Howells (q.v.), he had a similar philosophy of realism and sociological sympathy, but he lacked the Howells genius for sustaining a narrative interest.

GLASGOW, Ellen [Anderson Gholson] (1874-1945), was the daughter of a prosperous manager of a Richmond ironworks. Being a delicate child, she was educated at home. Her first novel, *The Descendant* (1897), was published anonymously. Most of her novels reflect her revolt against the mores of Southern gentility and her deep interest in social problems. *Virginia* (1913) and *Life and Gabriella* (1916) take up the problem of the southern woman in an industrial society. *Barren Ground* (1925) portrays a woman as the only source of strength to a decaying, rural community, and her *Vein of Iron* (1935) is again the feminine ore. Her three comedies of manners, somewhat reminiscent of the Augustan period, were intended to fill her prescription for the ailments of the South, a compound of "blood and irony." *The Romantic Comedians* (1926) portrays senility plucking spring flowers. *They Stooped to Folly* (1929) casts her inevitable weak man in the role of a tennis ball for a game of doubles between a mother-and-daughter team and a husband-secretary partnership. When the game is over the ball is no longer useful to either side. *The Sheltered Life* (1932) has more tenderness and less comedy. *In This Our Life* (1941) was awarded the Pulitzer Prize. *A Certain Measure* (1943) follows the Henry James precedent of providing retrospective prefaces for the author's novels.

GLASPELL, Susan (1882-1948), was born in Davenport, Iowa, graduated from Drake University, and reported for the Des Moines *Daily News* until her magazine stories were sufficiently successful to permit a trip to Paris for a year. In 1913 she married George Cram Cook, and together they founded the Provincetown Players, for which she wrote one-act plays, and also served as actor and producer. Later they continued their little-theater productions in the Playwrights' Theater, a converted stable in Greenwich Village. Her play about the effect of a deceased Iowa poetess on her immediate family, *Alison's House* (1930), was awarded the Pulitzer Prize, and is assumed to be based on the posthumous career of Emily Dickinson. In addition to innumerable plays, Miss Glaspell published 9 novels and a collection of short stories.

GODFREY, Thomas (1736-1763), born in Philadelphia, was the author of *The Prince of Parthia* (written 1759, probably first acted 1767), the first play by a native American to be produced on the professional stage. This tragedy of the remote Orient, written in blank verse according to the Elizabethan standards, is a tale of passion and violence. Godfrey also wrote many poems cut to British patterns. He died in Wilmington, North Carolina, where he had lived for several years and where he had completed his famous tragedy. His two volumes are a book of poetry, *The Court of Fancy* (1762), and *Juvenile Poems on Various Subjects, with The Prince of Parthia, a Tragedy* (1765).

GRAYSON, William J[ohn] (1788-1863), a South Carolina lawyer and poet, wrote *The Hireling and the Slave* (1854), a propaganda poem which argued that the Negro slave in the South had a far superior life to that of the Northern wage-slave.

Great Awakening, The (c. 1735-50) is a term used to identify an 18th-century religious revival in New England, led and reported mainly by Jonathan Edwards (q.v.). Edwards reported (in a letter to the Rev. Mr. Prince of Boston, Dec. 12, 1743) that the town of Northampton "has been in no measure, so free of vice . . . , for any long time together for sixty years, as it has been these nine years past." The new religious fervor was popular and emotional, and therefore quite different from the stern Calvinism of the previous century. "It was a very frequent thing, to see an house full of out-cries, fainting, convulsions, and such like, both with distress and also with admiration and joy." Peter Cartwright (1785-1872) later reported a similar revival, beginning in 1801 in the Cumberland country of Kentucky and Tennessee, called the *jerks*. "I have seen more than five hundred persons jerking at one time in my large congregations." This emotional reaction to conversion still persists in contemporary "revival meetings," especially in the South and Southwest. Edwards credited the New England revival in part to an act of Divine providence on March 13, 1737. An old and decayed meetinghouse collapsed on the congregation, but Providence disposed "the motions of every piece of timber, and the precise place of safety where everyone should sit and fall," so that none were killed. Powerless to prevent the disaster, Providence appeared capable of controlling and mitigating the effect, thus providing a practical argument of its power over the lives of men.

GREELEY, Horace (1811-72), the son of poor New Hampshire farm laborers, was apprenticed to a printer at 14, and at 22 became a partner in a New York printing business. The following year he founded the *New Yorker,* a critical weekly, which he merged with the *Tribune* in 1841. For the next three decades he championed an inconsistent variety of liberal and conservative causes (i.e., both high tariffs and socialism). In 1872 he was defeated by Grant for the presidency. He made no direct contribution to American literature, but his stimulating influence on national ideas and his employment of many famous writers helped to establish him as a legendary figure, now remembered as the author of one sentence: "Go West, young man!"

GREEN, Asa (1789-c. 1837), was trained as a physician but devoted most of his career to the writing of humorous novels. *Travels in America by George Fibbleton, Exbarber to His Majesty the King of Great Britain* (1835) is a burlesque of the observations of Mrs. Trollope. *The Perils of Pearl Street* (1834) is a more realistic, semi-autobiographical novel of economic problems and high finance. *A Yankee among the Nullifiers* (1833) records his visit to South Carolina. His works are a valuable source for study of the upper middle class before the Civil War.

GREEN, Paul [Eliot] (1894-   ), though the author of short stories, poetry, motion picture scenarios, essays, and novels, is primarily known for his work in folk-plays, the experimental theatre, and "symphonic drama," the last a term he originated. Born near Lillington, North Carolina, he began to write one act plays of the "folk" in the classes of Professor Frederick H. Koch while a student at the State University. In 1927 he won the Pulitzer Prize for his first full-length play, *In Abraham's Bosom*, which told of the defeat by both races of a mulatto schoolmaster in the South. *The Field God* (1927) deals with the conflict of a man's personal religion with the mores of a backwoods Southern neighborhood. *The House of Connelly* (1931) concerns the downfall of an aristocratic Southern house; it is generally considered his most successful Broadway play. In several experimental dramas, Green attempted new forms, new methods, new language, and new staging: *Tread the Green Grass* (1932) is described as "a folk fantasy with music and dumb show"; *Roll, Sweet Chariot* (1934, a revision of *Potter's Field*) is an intermissionless play of the composite life of a Negro community; and *Shroud My Body Down* (1934) is "A Folk Dream . . . To be produced by living actors trained in the manner of marionettes and with masks when so specified." Both *Johnny Johnson* (1936), a musical satire on war, and *Native Son* (1941), a dramatic collaboration with novelist Richard Wright, had satisfactory Broadway runs. The "symphonic dramas," employing ballet, chorus, narrator, and other adjuncts of the theatre, are written for annual summer presentation on vast outdoor stages.

Based on some chapter in American history, the most successful have been *The Lost Colony* (1937), about Sir Walter Raleigh's settlers of 1587 on the North Carolina coast; *The Common Glory* (1947), about Thomas Jefferson at Williamsburg; and *Faith of Our Fathers* (1950), concerning George Washington. Most of Green's work is fused with his belief in the "American dream," which, though only partially realized, must make men optimistic. He lives in Chapel Hill.

GREGORY, Horace [Victor] (1898-), was tutored at home, studied painting in Milwaukee academies, graduated from the University of Wisconsin, and became a free-lance writer in New York. His poems were published in *Vanity Fair, The Nation,* and *Poetry,* and his first book of poems, *Chelsea Rooming House,* was published in 1930, followed by *No Retreat* (1933), *Chorus for Survival* (1935), and *Poems: 1930-1940* (1941), a selection. His poems are urban in setting, conversational in tone, and dramatic in structure. He has also written a translation of Catullus, a critical study of D. H. Lawrence—*The Shield of Achilles* (1944), and numerous essays on both poetry and art. With his wife, Marya Zaturenska, he wrote the *History of American Poetry 1900-1940* (1946).

GUITERMAN, Arthur (1871-1943), New York journalist, wrote several volumes of humorous verse, from *The Laughing Muse* (1915) to *Brave Laughter* in 1943. He wrote the libretto for the opera by Walter Damrosch, *The Man Without a Country* (1937).

(95)

GUTHRIE, A[lbert] B[ertram] Jr. (1901-), grew up in Montana, and after graduating from the state university became a newspaper editor in Lexington, Kentucky. He began writing in 1944, while on a Nieman fellowship at Harvard, and resigned his editorial position in 1947 after the success of his first novel, *The Big Sky* (1947), a story of the fur trade in the early West. *The Way West* (1950) relates the trip of a group of pioneers traveling from Missouri to Oregon in 1846.

H

HALE, Edward Everett (1822-1909), a Boston Unitarian clergyman, wrote poems, stories, and essays, mostly didactic in nature, but is best remembered for *The Man Without a Country* (1865).

HALE, Sarah Josepha [Buell] (1788-1879), editor of the Boston *Ladies Magazine* (1828-37) and *Godey's Lady's Book* (1837-77). She sponsored child welfare programs and feminine education, and wrote plays, poetry, and novels. Her only remembered work has moved into the oral tradition, so that her name is seldom associated with "Mary Had a Little Lamb," published in *Poems for Our Children* (1830).

HALL, Clement (d. 1759), a native of England, was living in North Carolina as early as 1729. Ordained abroad in 1743, he returned to the colony as a Church of England missionary and later served in the northern part of the colony, becoming rector of St. Paul's, Edenton, in 1757. *A Collection of Many Christian Experiences, Sentences, and Several Places of Scripture Improved,* etc. (1753), the first non-legal work of a North Carolinian, is, though largely a compilation and interpretation, an excellent example of the early literature of religious devotion.

HALLECK, Fitz-Greene (1790-1867), was co-author, with J. R. Drake, of the "Croaker Papers," satiric verse lampoons published in the New York *Evening Post.* He also wrote *Fanny* (1819), a burlesque of the *nouveaux riches.* A friend of Dickens, Cooper, and Joseph and Louis Napoleon, a confidential secretary to John Jacob Astor, he never took his literary burlesques as seriously as his readers did, but considered himself a literary amateur. His collected *Works* were published in 1847.

HALLIBURTON, Richard (1900-39?), graduated from Princeton (1921) at a time when it was considered exceedingly glamorous for a young man to work his passage to Europe on a cattle boat, and young Halliburton promptly worked his passage to Europe on a cattle boat after using family influence to get himself hired. His family could have easily provided a first class passage, but that would not have been romantic. He continued working his way to every romantic part of the globe, and then wrote a very successful book, *The Royal*

Road to Romance, describing his adventures in swimming the Hellespont and bathing in the sacred pool of the Taj Mahal. He continued romancing and writing until 1939, when he sailed from China in a Chinese junk, headed for San Francisco. He never reached San Francisco.

HALPER, Albert (1904-   ), grew up on Chicago's West Side, and supported himself at a variety of odd jobs while studying accounting at Northwestern University and law in night school. His first novel, *Union Square* (1933), set among New York tenements, was a Literary Guild selection. His other novels, dealing with the aspirations and frustrations of the underprivileged, include *The Foundry* (1934), *Sons of the Fathers* (1940), *The Little People* (1942), and *Only an Inch from Glory* (1943).

HAMMERSTEIN, Oscar II (1895-   ), has written librettos and lyrics for the musical plays of Gershwin, Kern, and Richard Rodgers, including *Oklahoma* (1943) and *Carmen Jones* (1945).

HAMMETT, [Samuel] Dashiel (1894-), is one of the major authors of the mystery story novel, into which he injected certain qualities of realism and sophistication. His best known books are *The Maltese Falcon* (1930) and *The Thin Man* (1932).

HAMMETT, Samuel Adams (1816-65), was born in Connecticut and graduated from the University of the City of New York. He spent ten years in the Southwest, and reported his adventures with unusual veracity in *A Stray Yankee in Texas* (1853), *The Wonderful Adventures of Captain Priest* (1855), and *Piney Woods Tavern* (1858), using the pseudonym, Philip Paxton.

HARIOT, Thomas (1560-1621), wrote the first English book on what is now the United States. *A Brief and True Report of the New Found Land of Virginia* (1588) is a statistical survey of the products of the Roanoke Island area in North Carolina (then known as Virginia). A concluding section reports on the characteristics of the Indians. Hariot was scientist, surveyor, and historian of Sir Walter Raleigh's first colony (1585-1586). His treatise, in clear and precise Elizabethan English, was widely plagiarized by other writers on explorations, and thus its influence was extensive.

Harper's Magazine (1850-   ), originally *Harper's Monthly Magazine,* was founded by the publishing firm, Harper and Brothers. Under its first editor, Henry J. Raymond, it printed Dickens, Thackeray, Trollope, and Hardy. Later an increasing number of Americans were added to the list of contributors: Henry James, Howells, Melville, Garland, Sarah Orne Jewett, and Mark Twain. After 1900 an increasing amount of space was devoted to political and social issues. Frederick Lewis Allen was editor from 1941 to 1953, when he was succeeded by John Fischer.

HARRIS, Frank (1856-1931), was born in Ireland, and came to the United States in 1870, becoming an American citizen five years later. He held various jobs, including cattle punching, later reported in *My Reminiscences as a Cowboy* (1930). After study at the University of Kansas he was admitted to the state bar. Returning to England for additional study, he became editor of *The Fortnightly Review*, the *Saturday Review*, and *Vanity Fair*, and an intimate friend of Max Beerbohm, Shaw, H. G. Wells, and Oscar Wilde. He won considerable notoriety, not only for a tendency toward profanity in the company of great persons, but for his shocking and probably inaccurate *Oscar Wilde: His Life and Confessions* (2 vols., 1916) and *Contemporary Portraits* (5 vols., 1915-27). His own *My Life and Loves* (3 vols., 1923-7) was no less frank and scandalous. He wrote several novels, collections of short stories, and plays.

HARRIS, George Washington (1814-69), as a Tennessee jeweler's apprentice, constructed a model of the first Tennessee River steamboat, which attracted the attention of an influential citizen who gave him the opportunity to become a river boat captain. Meanwhile he wrote technical articles for the *Scientific American*, and later shifted to political tracts and humorous sketches. His single published book, *Sut Lovingood Yarns* (1867), is a precursor of Twain in the frontier comedy tradition, and more realistic than Twain in its amoral representation of vivid frontier dialogue that has the quality at times of folk poetry.

HARRIS, Joel Chandler (1848-1908), first became acquainted with Negro legends when he was apprenticed as a typesetter with the newspaper, *Countryman*, published on a Georgia plantation. After the Civil War he found newspaper work in Macon, New Orleans, Savannah, and finally, with the Atlanta *Constitution*, where he began to publish a column featuring his Uncle Remus tales (1879). These were later collected as *Uncle Remus: His Songs and His Sayings* (1880), followed by a series of additional volumes until 1918. These stories remain the greatest of Negro folk literature, rich in humor, authentic in plantation dialect and attitudes, but with a human reference far broader than any mere sectional study.

HARRISON, Constance [Cary] (1843-1920), granddaughter of Lord Fairfax, was born in Virginia, and lived in Arlington, New York, and England. She wrote stories and novels of New York society and Virginia life, including *Belhaven Tales* (1892) and *A Bachelor Maid* (1894). *The Anglomaniacs* (1899) is a comic picture of American social climbers abroad, and with *Sweet Bells Out of Tune* (1892) she turned her comic attention to the English society.

HARRISON, James Albert (1848-1911), wrote popular books on Greece and Spain, a book of Creole stories, and a biography and study of George Washington, but his more serious work was in philology. He edited an Old-English edition of *Beowulf*, and *A Dictionary of Anglo-Saxon Poetry*.

HART, Moss (1904-), collaborated with George S. Kaufman in writing:

Once in a Lifetime (1930); *You Can't Take It With You* (1936; Pulitzer Prize, 1937); *I'd Rather Be Right* (1937); *The Man Who Came to Dinner* (1939); and *George Washington Slept Here* (1940). In addition to plays, he has written librettos for the revues of Irving Berlin.

HARTE, [Francis] Bret[t] (1836-1902), was born in Albany, New York, but his teaching father moved constantly looking for material success, and died in 1850. In 1854 Bret followed his mother to San Francisco, and became a tutor, an express messenger, a miner, a drugstore clerk, a printer, and finally assistant editor of the *Northern Californian*. Fired for condemning the assassination of Indians, he obtained a typesetting job with the San Francisco *Golden Era*, and soon was writing a column. His appointment as Secretary of the California Mint afforded him security, and leisure to edit the *Overland Monthly*, in which his famous stories were published, including "The Luck of Roaring Camp" and "The Outcasts of Poker Flat," as well as the comic ballad, "Plain Language from Truthful James." In 1867 his collection of poems, *The Lost Galleon*, was published locally, and also *Condensed Novels and Other Papers*. During this period he was a leader of California literary activities and a friend of Jessie Benton Frémont, Mark Twain, and James W. Gillis, the Truthful James of Harte's poem. Meanwhile his work had won enthusiastic admiration in the East, and in 1871 he left California, accepting an offer of $10,000 from the *Atlantic*

Monthly for the next year of his production. It amounted to four stories and five poems, disappointing in both quantity and quality. Thenceforth his career gradually declined. His lecture tours were not successful, and neither was his novel, *Gabriel Conroy* (1876), nor the play written with Twain, *Ah Sin* (1877). He spent his last years in London, supporting himself by hack writing, mainly in imitation of his earlier California work.

HAVIGHURST, Walter [Edwin] (1901-), grew up in the Mid-West, graduated from the University of Denver, and after taking graduate work at King's College, London, took his M. A. degree at Columbia. His novels include *Pier 17* (1935), about a seaman's strike on the West Coast; *The Quiet Shore* (1937), *The Winds of Spring* (1940), and *The Signature of Time* (1949). His historical prose includes a contribution to the "Rivers of America" series, *Upper Mississippi* (1937).

HAWTHORNE, Nathaniel (1804-1864), was born in Salem, Massachusetts, a descendant of a long line of Puritans and the great grandson of a judge at the Salem witchcraft trials who appears in Hawthorne's works as the founder of *The House of the Seven Gables* (1851), a novel that pictures the decadence of Puritanism. His father was lost at sea when he was four, and his mother introduced him to literature at an early age in the hope of diverting him from the maritime profession. A leg injury suffered at nine while playing ball further constricted him to intellectual pursuits. He attended Bowdoin, where Franklin Pierce was a classmate. He returned to Salem, determined to write, but his first novel, *Fanshawe* (1828), published anonymously at his own expense, was a failure. During the thirties many of his stories appeared, again anonymously, in *The Token*, and were collected as *Twice-Told Tales* (1837) after his college friend, Horatio Bridge, had provided the publisher a guaranty of $250. The book included many of his now famous stories: "Mr. Higginbotham's Catastrophe," "Dr. Heidegger's Experiment," "The Ambitious Guest," and "The Minister's Black Veil." In 1839 he accepted an appointment as "measurer of salt, coal, etc.," at the Boston Custom House, but resigned in two years and joined the Brook Farm Institute (q.v.), from which he also resigned six months later. In 1842 he married Sophia Peabody, and they settled in the Old Manse at Concord, where he wrote the stories collected in *Mosses from an Old Manse* (1846). He was Surveyor of the Port of Salem from 1846 to 1849, an experience he described in a prefatory essay to *The Scarlet Letter* (1850). He next moved to Lenox, Mass., where Melville was a neighbor and frequent caller. After another move, to West Newton, he settled in Concord (1852) and wrote a campaign biography for his college friend, General Franklin Pierce. His reward was the consulship at Liverpool, where he resided until 1857, consistently impatient with his duties and eager to return

to full-time writing. Before sailing for the United States in 1860, he toured the continent and stayed in Italy for two years, an experience that provided the source for *The Marble Faun* (1860). In May of 1864, he departed with Pierce for a vacation in the White Mountains, and while lodging near Plymouth, Hawthorne died in his sleep.

Hawthorne was unable to identify himself with the main currents of his contemporary life, and seemed always to be haunted by his Puritan ancestors, so that problems of guilt or moral and intellectual pride constantly reappear in his work. The story he wrote for the 1837 edition of *Twice-Told Tales*, "Endicott and the Red Cross," tormented him until he expanded the story, of the adulteress forced to wear the scarlet letter "A", into a novel, *The Scarlet Letter*. *The House of the Seven Gables* (1851) seems almost an attempt by the author to expiate the curse pronounced on his own family for the judgment of his great-grandfather at the Salem witchcraft trials, for the *House* is freed from an identical curse at the end of the tale. *The Blithedale Romance* (1852), a fictional portrait of Brook Farm, leaves the spectator sceptical of the possibilities for human progress. Like Melville, he was too deep, both in his philosophy and his psychology, for most of his contemporaries, but his reputation has steadily increased since the days when an obscure editor of the Brooklyn *Eagle* named Walt Whitman could complain in his editorial, "Shall Hawthorne get a paltry $75 for a two volume work?" (July 11, 1846).

HAY, John [Milton] (1838-1905), grew up in Indiana, graduated from Brown University, and studied law in an uncle's office at Springfield, Illinois. He served President Lincoln as assistant private secretary, and after the Civil War was secretary of legation at Paris, Vienna, and Madrid. His *Castilian Days* (1871) brought him temporary fame, which was considerably increased with the publication of *Pike County Ballads* (1871), dialect verses of the frontier. He deprecated their "absurd vogue," but his more serious work had much less literary appeal to future generations. *The Breadwinners* (1883) was a novel in defense of capitalism. With J. G. Nicolay he wrote *Abraham Lincoln: A History* (10 vols., 1890), a valuable source for the study of the Civil War president. Hay served as Ambassador to Great Britain (1897-8) and as Secretary of State under McKinley and Roosevelt.

HAYDN, Hiram (1907-), novelist, teacher, and editor of *The American Scholar*, took his Ph. D. at Columbia. His novels offer studies of the conflicts between dreams and reality, and include *By Nature Free* (1943), *Manhattan Furlough* (1945), and *The Time is Noon* (1948), a portrait of American life through the search of six characters for happiness. His scholarly study of *The Counter Renaissance* (1950) casts Machiavelli and Montaigne in the title rolls.

HAYNE, Paul Hamilton (1830 - 86), a South Carolina poet, was trained for the law, but followed his major interest in founding and editing *Russell's Magazine*

(1857-60). His *Poems* (1855), *Sonnets and Other Poems* (1857), and *Avolio* (1860) were praised by the Northern critics for their tunefulness and delicate beauty. A member of a wealthy family in the genteel tradition, he was impoverished by the war, but continued to write in his shack at "Copse Hill" for the best Northern magazines. *Legends and Lyrics* (1872) contains his most mature work.

HEARN, Lafcadio (1850 - 1904), was born on an Ionian island, the son of a surgeon-major in the British army and a Greek beauty. At 6 he was entrusted to a Dublin aunt and brought up in Catholic schools, but after his siege of rebellion she sent him to America, and following years of poverty he became a reporter on the Cincinnati *Enquirer*. After an entanglement with a mulatto woman cost him his job, he went to New Orleans, where he achieved considerable success in journalism, and investigated Creole culture with George Washington Cable. His first book, *One of Cleopatra's Nights* (1882) contained translations from Gautier, and was followed by fragments of exotic literature collected as *Stray Leaves from Strange Literature* (1884). *Gombo Zhèbes* (1885) contains Creole proverbs, and *Some Chinese Ghosts* (1887) presents highly polished Oriental legends. *Chita: A Memory of Last Island* (1886) reports a tidal disaster he witnessed in the Mississippi Gulf. A two-year residence in Martinique provided material for his novel about a beautiful slave girl, *Youma* (1890). In 1890, after an unpleasant visit to New York that resulted in quarrels with friends, he sailed for Japan, where he lived out his life, marrying into a prominent Samurai family, occupying the chair of English literature at the University of Tokio, and becoming a Japanese citizen. Here he wrote 12 books of stories or sketches about Japan, including *Kokoro* (1896), *Glimpses of Unfamiliar Japan* (1894), and *Japan: An Attempt at Interpretation* (1904).

HECHT, Ben (1894-), refused to go to college after graduating from high school in Racine, Wisconsin, and became an acrobat with a touring road show. While still in his teens he started newspaper work in Chicago, and later moved in the literary circles there which included Carl Sandburg and Sherwood Anderson. With Maxwell Bodenheim he wrote *The Master Poisoner* (1918) and several other one-act plays. During the early twenties he published more than a dozen novels, short story collections, and plays, but he became nationally famous with the production of *The Front Page* in 1928, an irreverent portrait of newspaper life, written in collaboration with Charles MacArthur. He has written many scenarios, including the experimental *Crime Without Passion* and *The Scoundrel*, again with MacArthur.

HELLMAN, Lillian (1905-), New York dramatist and scenarist, was educated at Columbia and New York Universities. Her play based on a scandal of early 19th-century Scotland, *The Children's Hour*, (1934) played for 691 performances in New York. The story presents a malicious child who lies about the proprietors of her boarding school in such a way as to suggest to her gossiping elders that they are Lesbians. *The Little Foxes* (1939) portrays decadence in a reactionary Southern family. Her other plays are *Days to Come* (1936), *Watch on the Rhine* (1941), and *The Searching Wind* (1944).

HELPER, Hinton Rowan (1829-1909), born in Davie County, North Carolina, wrote *The Impending Crisis of the South: How to Meet It* (1857), the most damning book on slavery written by a Southerner. Unsympathetic with the Negro, it blamed slavery for Southern economic and social backwardness. It was banned in the South, but read in the North for its antislavary propaganda second only to *Uncle Tom's Cabin*. John Sherman, Republican leader in Congress, issued a *Compendium* to it. Among Helper's other controversial works are *Land of Gold: Reality Versus Fiction* (1855), which was highly critical of California; *The Negroes in Negroland* (1868) and similar books advocating the extinction of the black race; and *The Three Americas Railway* (1881), one of his many volumes proposing the construction of a railroad from Hudson Bay to the Strait of Magellan. Helper served (1865-66) as consul at Buenos Aires; he was later a lawyer in Washington.

HEMINGWAY, Ernest [Miller] (1898-), the son of a physician in Oak Park, Ill., accompanied his father on frequent hunting trips to northern Michigan, and was prominent in high school football and boxing. After reporting for the Kansas

City *Star,* he served with an American ambulance unit in France, then joined the Italian infantry, and was wounded while fighting on the Italian front. He was decorated with the Medaglia d'Argento al Valore Militare and the Croce di Guerra. After the war he married Hadley Richardson and became European correspondent for the Toronto *Star,* reporting post-war disturbances in the Near East, and the Greek revolution. In Paris he came under the influence of Gertrude Stein, Sherwood Anderson, and Ezra Pound, whom he tried to teach boxing. His first signed literary work appears in *Poetry* (Jan., 1923). His early story collections were published in Paris: *Three Stories and Ten Poems* (1923) and *in our time* (1924). These stories set the pattern of most of his future work, presenting cynical sophisticates or noble primitives, each speaking his own type of tough language. *The Sun Also Rises* (1926) presents the "lost generation" theme of his stories in a novel about war-wounded American and British expatriates seeking a meaning or at least a tenable attitude during a fiesta in Spain, and a bullfighter is the inconspicuous hero. *The Torrents of Spring,* published earlier the same year, burlesques the style of Sherwood Anderson, and is important only as a study of the Anderson influence on Hemingway. *Men Without Women* (1927) contains many of his best stories, including "The Undefeated," a tragedy of the bullfighter past his prime, and "The Killers." *Death in the Afternoon* (1932) is a book-length essay on the aesthetics and philosophy of bullfighting, while *Green Hills of Africa*

(1935) is a somewhat similar study of big-game hunting. *A Farewell to Arms* (1929) is a tender love tragedy about a man and woman engulfed by the unreasonable demands of war. *To Have and Have Not* (1937) contrasts a Key West rum-runner and his kind with the yacht owners, writers and artists. Hemingway here seems to be affected by the neo-Naturalism of the thirties, but his contrast is as much between primitive and sophisticated man as between economic classes. *For Whom the Bell Tolls* (1940), about an American fighting with the Loyalists in the Spanish civil war, avoids the propaganda cliches because it is more about a man and men than about ideologies. *The Fifth Column and the First Forty-Nine Stories* (1938) contains a three-act play describing communist and fascist counter espionage in Madrid, and his most famous story, "The Snows of Kilimanjaro," picturing a hunter dying in Africa and haunted in his stream of consciousness by the futility of his glamorous life. *Across the River and into the Trees* (1950), one of his least successful books, perhaps fulfilled his need to get a lot of words off his chest about World War II. In 1927 Hemingway married Pauline Pfeiffer, of *Vogue* magazine, Paris staff, and they lived in Key West for the next ten years, interrupted by African hunting trips. In 1935 he was still able to put on a good show boxing Tom Heeney, the British heavyweight champion. He took a great interest in the Spanish war, contributed ambulances to the Loyalist side, and reported the fight for the American press. Early in World War II (1942-44) he cruised the Caribbean in his fishing yacht Pilar, to provide the Navy with intelligence in the submarine warfare. In 1944 he served with the Royal Air Force as correspondent, and later, with the United States army infantry, he organized his own force of 200 "papa's irregulars," which he sometimes led up to 60 miles ahead of the front line. In 1940 he had married the novelist, Martha Gelhorn, but they were later divorced. In 1946 he married Martha Welsh, a *Time* magazine correspondent, in Cuba, where he maintains an estate of fifteen acres. Havana and the Cuban waters where he has often fished are also the scene of his most recent work, *The Old Man and the Sea* (1952) a novella first published in an issue of *Life* magazine. It is a powerful story of an ancient fisherman's battle to catch and bring in to shore a great Marlin. With an almost perfect structure, with the intensity and unity of action of a short story, it is perhaps his best technical performance. Hemingway has achieved enormous success in his own time, and is one of the few authors of distinction whose works have often been made into motion pictures, perhaps because he is one of the few who consistently choose a dramatic setting. His style, barren of cosmetics, has changed the prose of American fiction, yet none of his disciples has matched his quality. His short, staccato sentences are characteristic of the "tough" school of composition. More than any other novelist he represents the complete reaction against the Victorian style of writing. In 1954 he was awarded the Nobel Prize for literature in recognition of his development of a strong, masculine style.

HENDERSON, Alice Corbin (1881-   ), editor of *Poetry* Magazine from 1912 to 1916, when she moved to New Mexico. She wrote several volumes of poetry. *The Turquoise Trail* (1928) is her anthology of New Mexico poets.

HENRY, O. (1862-1910), was the penname used by William Sydney Porter, America's most popular and widely read short story writer. At the height of his career he was writing stories one a week for the New York newspapers and periodicals. Noted for his careful plotting and surprise endings, he was nevertheless a master of the vignette, of the ironic and occasionally humorous coincidences which shape situations. Most of his characters were drawn from the lives of everyday Americans, especially those in the cities, though he dealt also with the South and West, as well as Central and South America. He possessed a genuine sympathy for the underdogs of life and could picture their woes and their happiness in melting prose. The approximately 300 stories are contained in *Cabbages and Kings* (1904),*The Four Million* (1906), *Heart of the West* (1907), *The Trimmed Lamp* (1907), *The Gentle Grafter* (1908), *The Voice of the City* (1908), *Options* (1909), *Roads of Destiny* (1909), *Whirligigs* (1910), *Strictly Business* (1910), *Let Me Feel Your Pulse* (1910), *Sixes and Sevens* (1911), *Rolling Stones* (1913), *Waifs and Strays* (1917) and *Postscripts* (1923). Born in Greensboro,

North Carolina, Porter attended the private school of his aunt, "Miss Lina"; then, after working in his uncle's drugstore, he went to Texas, where he wrote for a Houston newspaper and worked in an Austin bank until he was accused of embezzlement and escaped to Honduras. Upon hearing of his wife's illness, he returned, and was apprehended and sentenced to prison, where he started writing short stories. Upon his release, he settled in New York. He is buried in Asheville, North Carolina, not far from the grave of Thomas Wolfe. With his second wife, he had returned to his native state in his declining days. Though unauthenticated, it is said Porter adopted the pseudonym O. Henry both because he wished to conceal his prison identity and because it was a name easily remembered.

HERGESHEIMER, Joseph (1880-1954), grew up in Philadelphia, and studied painting at the Pennsylvania Academy of Fine Arts, and in Italy until he had spent his inheritance. He returned to the United States and put in a long apprenticeship as a writer, living in the Virginia mountains. He was a frequent contributor to the *Saturday Evening Post,* and as a professional writer produced a number of popular novels, in addition to a few of distinction. *The Three Black Pennys* (1917) portrays the Pennsylvania iron industry and the changing fortunes of the family behind a foundry during three generations. *Java Head* (1919) is a historical study of miscegenation in New England.

HERNE, James A. (1839-1901), born James Ahern, a New York actor and dramatist, wrote *Hearts of Oak* (1879) with David Belasco, and with the beautiful Mrs. Herne he acted in the San Francisco production. This was a pioneer play in the break from stock characters and situations. *Shore Acres* (1888), written under the influence of Ibsen and Tolstoy, and with the encouragement of Hamlin Garland and William Dean Howells, did not get a production until 1893, and then became a long success. *The Reverend Griffith Davenport* (1899) was highly praised by William Archer, but was not a success, nor was *Sag Harbor* (1899). He had a considerable influence, however, on early 20th-century American drama, as a pioneer of outspoken social drama. (See also DRAMA).

HERRICK, Robert (1868 - 1938), attended the Cambridge Latin School and graduated from Harvard, where he was an editor of the *Advocate* and the *Harvard Monthly,* and a friend of William Vaughn Moody, Robert Morse Lovett, and George Santayana. He taught at the University of Chicago for thirty years (1893-1923), giving most of his spare time to his career as a writer. Many of his novels were concerned with the conflict between ambition for material success and personal ideals, as in *The Common Lot* (1904), concerning an architect who builds cheap tenements that crumble on the tenants. *The Master of the Inn* (1908) deplores the effect of city-dwelling on the mind and body.

HERSEY, John [Richard] (1914-   ), the son of missionaries, was born in Tientsin, China, and after a primary education there, graduated from Yale in 1936. He was a secretary for Sinclair Lewis in 1937, and later became a *Time* magazine correspondent in the Orient, where he obtained background materials for *Men on Bataan* (1942) and *Into the Valley* (1943). His first novel, *A Bell for Adano* (1944) portrays an Italian-American major in his efforts to reconstruct a village in Sicily. It won the Pulitzer Prize in 1945. *Hiroshima* (1946), an objective report of the atom bombings, was first published in *The New Yorker,* occupying the entire issue.

HEWAT, Alexander (c. 1745-1829), a Scottish clergyman, emigrated to South Carolina in 1763. His *An Historical Account of the Rise and Progress of the Colonies of South Carolina and Georgia* (2 vols., 1779) is considered unreliable in its references to the Revolutionary War because of his Loyalist sympathies.

HEYWARD, Du Bose (1885-1940), was born into an aristocratic Charleston family impoverished by the effects of the Civil War, and his father died when he was two, so that he was forced to sell papers at nine and to leave school at 14 for full employment, first with a hardware store, then with a steamship line, and as a cotton checker. Later he entered the insurance busines. With Hervey Allen he published *Carolina Chansons* (1922), followed by his own collection of poems, *Skylines and Horizons* (1924). In 1923 he married Dorothy Hartzell Kuhns, a playwright, who helped him to dramatize his successful novel, *Porgy,* (1925), a story of Negroes based on his acquaintance with that race while working on the Charleston docks. Produced by the Theatre Guild with an all Negro cast, the play won the 1927 Pulitzer Prize, and was later produced as the opera, *Porgy and Bess* by George Gershwin (1935). He also published another book of poems, *Jasbo Brown* (1931), and the novels, *Peter Ashley* (1932) and *Star Spangled Virgin* (1939).

HICKS, Granville (1901-   ), grew up in New Hampshire, and graduated from Harvard (1923) and Harvard Divinity School (1925). While teaching Biblical literature and English at Smith College, he contributed articles to the *American Mercury, Forum, Nation,* and *New Republic,* and wrote *Eight Ways of Looking at Christianity* (1926). In 1929 he took his M. A. degree at Harvard, and in the thirties, became identified with the Marxist movement to establish economic determinism as a criterion of value in literary criticism. He became literary editor of *The New Masses,* and wrote a Marxist interpretation of American literature, *The Great Tradition* (1933), which reappraised each writer solely in terms of his contribution to the analysis of our economic problems. *Figures of Transition* (1939) similarly appraises British literature in the latter 19th century. In 1939 he resigned from the Communist Party. His novel, *Only One Storm* (1942), tells the story of a man engulfed by conflicting intellectual dogmas.

HIGGINSON, Thomas Wentworth (1823-1911), grew up in Cambridge, attended Harvard and the Harvard Divinity School, and became a Unitarian minister. Later he was a leader of the radical abolition movement, and worked for the assistance of fugitive slaves. During the Civil War he led the first regiment of Negro soldiers, with the rank of colonel, and was wounded in action. He described this career in *Army Life in a Black Regiment* (1870). After the war he retired and became a prolific writer. In addition to the novel, *Malbone* (1869), he wrote sketches and biographies of Longfellow, Whittier, Margaret Fuller, and himself. A conservative in letters, he sided with the English philologists in the dispute over "Americanisms," and endeavored to correct the virtues in Emily Dickinson's poems.

HILLHOUSE, James Abraham (1789-1841), was highly praised for his conventional poems and verse dramas: *Percy's Masque* (1819), *The Judgment* (1821), *Hadad* (1825), and *Demetria* (1839). His work had eloquence and dignity which exceeded that of many of the conventional poems of his day.

HILLYER, Robert [Silliman] (1895-), graduated from Harvard in 1917, served with an ambulance unit in France and, as a lieutenant, with the AEF. He studied at the University of Copenhagen on a fellowship, and later was appointed Boylston Professor and Phi Beta Kappa poet at Harvard. His first of many volumes of poetry was *Sonnets and Other Lyrics* (1917), and his *Collected Verse* (1933),

was awarded the Pulitzer Prize. He has also written two novels, *Riverhead* (1932) and *My Heart for Hostage* (1942), as well as collections of critical essays: *Some Roots of English Poetry* (1933) and *First Principles of Verse* (1938). Recently he served as champion of the "old criticism" in his articles, published in the *Saturday Review of Literature,* which questioned the award of the 1949 Bollingen Prize to Ezra Pound.

HIRST, Henry Beck (1817-74), a Philadelphia lawyer whose poems in *The Coming of the Mammoth* (1845), etc., were greatly admired in his day. A friend of Poe for a time, he claimed to be the true author of "The Raven." He died in the Blockley Almshouse where he had been confined for insanity.

HOFFMAN, Charles Fenno (1806-84), New York lawyer, surveyor, editor, and novelist, won temporary fame for his successful novel, *Greyslaer* (1840), based on a Kentucky murder case, and later dramatized. He also wrote musical lyrics in praise of nature: *The Vigil of Faith* (1842), *The Echo* (1844), and *Love's Calendar* (1847). He lost his sanity in his middle age.

HOLLAND, Edwin Clifford (c. 1794-1824), South Carolina poet who is considered to mark the beginning of romantic poetry in that state. He wrote *Odes* (1813) and a dramatization of Byron's *Corsair* (1818).

HOLLAND, Josiah Gilbert (1819-81), Massachusetts author who wrote didactic and sentimental verse, novels, and miscellaneous prose, sometimes as "Timothy Titcomb." His major distinction was in achieving an incredible popularity with incredibly poor work.

HOLMES, Mary Jane [Hawes] (1825-1907), wrote 39 stereotyped novels (1854-1905) that sold more than 2,000,000 copies.

HOLMES, Oliver Wendell (1809 - 94), the son of a Cambridge minister, was reared in the tradition of New England gentility, and became interested in classical literature at Harvard. While attending law school he wrote the popular poem, "Old Ironsides" (1830), a protest against the proposed destruction of the fighting ship, "Constitution." In 1831 he transferred his study to medicine, at a private school, at Harvard, and for two years in Europe. As a general practitioner his practice was limited by his reputation as a poet and wit, and he gradually shifted to the academic aspects of medicine, serving as professor of anatomy at Dartmouth (1838-40), Dean of the Harvard Medical School, and Parkman Professor of Anatomy and Physiology at Harvard (1847-1882). Meanwhile he published lectures in opposition to homeopathy and on the nature of fevers. His reputation for lectures that were both accurate and witty provoked invitations to lecture before lyceums. He joined Lowell, Howells, and Longfellow in the "Saturday Club." He named the *Atlantic Monthly*, and his series of humorous conversational essays concerning *The Autocrat of the Breakfast-Table* (1858) are

credited with helping the new magazine through its first year of publication. As a militant Unitarian he attacked Calvinism in "The Deacon's Masterpiece" (1858). Of his three novels, *Elsie Venner* (1861) is the best remembered. He published five volumes of his poems, biographies of John Lothrop Motley and Emerson, essays, and travel sketches. One of his sons, Oliver Wendell Holmes, Jr. (1841-1935), was a Justice of the United States Supreme Court for many years.

HOOKER, Thomas (1586-1647), English-born Congregationalist who immigrated to Massachusetts, later opposed Governor Winthrop's magisterial autocracy, and took his congregation of a hundred families to found the Connecticut Colony. In *A Survey of the Summe of Church Discipline* (1648) he asserted the right and power of the individual to deal directly with God.

HOOPER, Johnson Jones (1815-1862), Southern frontier humorist, in *Some Adventures of Captain Simon Suggs* (1845) created a fascinating backwoods scoundrel whose rowdy conduct and spicy dialect influenced the writings of Mark Twain and Thackeray. The picaresque Simon Suggs episodes first appeared in newspapers and were widely copied. Born in Wilmington, North Carolina, Hooper rose to prominence as a Secessionist in Alabama and served as secretary of the Confederate Congress.

HOPKINS, Lemuel (1750-1801), a Connecticut physician and member of the "Connecticut Wits" (q.v.), wrote humorous, satirical verse, as "Epitaph on a Patient Killed by a Cancer Quack." He was one of the most advanced and distinguished physicians of his time.

HOPKINSON, Francis (1737-91), Philadelphia lawyer, musician, merchant, satirical essayist, and poet, was the first student and first graduate of the Academy of Philadelphia (now University of Pennsylvania). He wrote the first song published in America, was the first judge of the United States Court in Pennsylvania, and a signer of the Declaration of Independence. As an artist he designed the seal of New Jersey and helped to design the American flag. He wrote lyric, religious, and satirical poems such as "The Battle of the Kegs".

HOPWOOD, Avery (1882-1928), was a graduate of the University of Michigan which institution has been endowed to issue prizes in creative writing called the Avery Hopwood awards. He was a New York journalist and successful playwright. His works include *The Gold Diggers* (1919); *The Bat* (1920) written with Mary Roberts Rinehart; and *Getting Gertie's Garter* (1921), written with Wilson Collison.

HORGAN, Paul (1903-), grew up in Buffalo, New York, and attended the New Mexico Military Institute, where he later became a member of the faculty. He studied at the Eastman School of Music and worked in stage productions before becoming a writer. He was awarded the Harper prize for *The Fault of Angels* (1933), concerning the conductor of an opera company and his troublesome

spouse. *No Quarter Given* (1935) pictures a composer dying in Santa Fe, with his past recalled in flashbacks. *The Habit of Empire* (1939) is set in 16th-century New Mexico, and describes the Spanish colonization. *Great River* (2 vols., 1954) is a superb study of the Rio Grande that combines accurate scholarship with an imaginative presentation. Mr. Horgan served in World War II with the rank of lieutenant colonel.

HORTON, George Moses (1797?-1883?), Negro slave poet who could not read and write until after his first book was published, and never really mastered the skills, was born on an eastern North Carolina plantation. At six, he moved with his master to a farm ten miles from the University of North Carolina and was soon reciting to himself original verses he had composed on the framework of the old Methodist hymns. While peddling farm produce at Chapel Hill, he attracted the attention of the students, who served as his amanuenses and paid him for love poems they sent to their girls back home. Horton secured a job as porter at the university, paying his master partially from his literary profits. In 1865 he followed a Union captain to Philadelphia, where he lived his remaining days. *The Hope of Liberty* (1829) was twice reprinted in the North as abolitionist propaganda; *Poetical Works* (1845) and *Naked Genius* (1865) are his other volumes. Horton's inventiveness and his variety of subject matter are quite remarkable, and his poetic gift is amazing, even discounting his educational handicap.

Hound and Horn (1927-34), founded by Lincoln Kirstein and Varian Fry, was a distinguished literary journal that published the work of Kenneth Burke, Allen Tate, Ezra Pound, T. S. Eliot, and Gertrude Stein.

HOVEY, Richard (1864-1900), son of a Civil War General and educator, graduated from Dartmouth College where he wrote most of the songs still sung there. He studied art and theology, and was an actor for several years. While in Europe (1891-2) he was influenced by the French Symbolists, Verlaine, Mallarmé, and Maeterlinck, whose plays he translated. With the Canadian poet, Bliss Carman, he wrote *Songs from Vagabondia* (1894) followed by "More" (1896) and "Last" (1901) songs, proclaiming the joys of youth on the open road. His projected cycle of poetic dramas on the Arthurian legend was not completed, although *The Holy Grail* (1907) was published posthumously.

HOWARD, Bronson (1842-1908), was a newspaper man in Detroit and later in New York until his play, *Saratoga* (1870), established him as one of the very successful playwrights of his day. He wrote twelve other plays, including *Young Mrs. Winthrop* (1882), *The Henrietta* (1887), and *Shenandoah* (1888), his most popular work. His plays were equally popular in London, where he maintained a second residence.

HOWARD, Sidney [Coe] (1891-1939), attended the University of California, and after graduation studied at Harvard with G. P. Baker (q. v.). In the first war he commanded an aviation combat squadron. On his return he worked as a magazine writer and wrote and adapted plays. His first success, *They Knew What They Wanted* (1924) won the Pulitzer Prize. Of his many other plays, the best known are *The Silver Cord* (1926), presenting a possessive mother in battle to hold her son; *The Late Christopher Bean* (1932), a French play transferred to New England, and *Dodsworth* (1934) dramatized with Sinclair Lewis. He also wrote the scenarios for a number of motion pictures. In 1939 he was killed by a tractor accident on his avocational farm.

HOWE, E[dgar] W[atson] (1853-1937), owned and edited the *Daily Globe* in Atchison, Kansas (1877-1911) and *E. W. Howe's Monthly* (1911-37). When his novel, *The Story of a Country Town* (1883), was rejected by the major publishers, he printed it himself. It is now recognized as a pioneering work in Naturalistic American fiction, combining a powerful drama with realistic delineation of small-town people in the Middle-West. He was known for his aphorisms as "the Sage of Potato Hill." His other works include: *The Confessions of John Whitlock* (1891), *Country Town Sayings* (1911), *Ventures in Common Sense* (1919), and his autobiography, *Plain People* (1929).

HOWE, Julia Ward (1819-1910), poet, biographer, editor, and lecturer, was the daughter of a prominent New York banker. With her husband, Dr. Samuel G. Howe, she edited the Boston *Com-*

monwealth, an abolitionist paper, and their Boston home became a center for the antislavery movement. She became famous in 1862 with the publication in the *Atlantic Monthly* of her poem, "The Battle-Hymn of the Republic," and was the first woman elected to the American Academy of Arts and Sciences. Two early plays were unsuccessful, and her five volumes of poetry have no great distinction, but her penetrating, critical mind served her well in such prose works as *Life of Margaret Fuller* (1883), *Sex and Education* (1874), and *Modern Society* (1881). After the death of her beloved husband she threw herself into several reform movements, including woman suffrage and prison reform. Her homes in New York and Newport became salons for literary gatherings. Laura E. Richards and Maud Howe Elliott, her daughters, both won distinction in literature.

HOWELLS, William Dean (1837-1920), was born in Martin's Ferry, Ohio, the son of a Welsh printer of Swedenborgian philosophies. At nine he was setting type in his father's shop, and although he had no formal education in the following years, he educated himself by constant reading and the study of languages. This period is described in *My Year in a Log Cabin* (1893) and *A Boy's Town* (1890). From 1856 to 1860 he was a compositor, reporter, and later news editor for the Columbus *Ohio State Journal.* Characteristically, he declined an attractive offer from Cincinnati because the position would include police reporting.

His first book was the volume of *Poems of Two Friends* (1860), J. J. Piatt being the second author. With the proceeds from his campaign biography of Lincoln, he visted New England, and met his Brahma gods, Lowell, Emerson, and Hawthorne. An additional reward for the biography was his appointment as United States consul in Venice (1861-5). During this period he met his future wife, Elinor Mead, in Paris. He also collected the material for two books: *Venetian Life* (1866); and *Italian Journeys* (1867), still charming and useful as one of the best guide books for Italy. Moreover, the Italian influence constantly reappeared in his work: *Modern Italian Poets* (1887) reveals a sound knowledge of the nation's language and literature; *A Foregone Conclusion* (1875) is a novel about Americans in Italy; and the shipboard romance of *The Lady of the Aroostook* (1879) reaches its climax in Venice, where he contrasts the two cultures in some of his most perceptive writing. After his return to America Howells worked briefly for the *New York Times* and the *Nation* before settling in Boston as sub-editor, and editor (1872-81), of the *Atlantic Monthly,* where he retained his midwestern equalitarianism in a curiously compatible merger with Brahmin cultural pretensions. His first novel, *Their Wedding Journey* (1872), was a transition between his travel sketches and the formal novel, presenting Isabel and Basil March on a honeymoon trip to Niagara Falls and Canada. The major dramatic conflict of the book develops in the question of whether Isabel will summon the courage to return over a footbridge from an island below the Falls, whereas a boat accident on their trip up the Hudson is presented as an incidental footnote to their adventures. Yet this work demonstrates better than any other the Howells genius for maintaining the reader's consistent interest without plot or any artificial device. Indeed, it is the more highly developed plot in such a novel as *A Modern Instance* (1881) that is apt to bore the modern reader, constructed as it is on a framework of 19th-century mores. *The Rise of Silas Lapham* (1885), often considered to be his masterpiece, is a sensitive portrait of a self-made man, and of his pride in achievement that goes before a fall, but unlike Bartley Hubbard of *A Modern Instance,* Lapham is felled by his subservience to ethics rather than by a lack of moral discrimination. Howells favored the realism of the average life rather than that of the depraved life. (Frank Norris (q. v.) later ridiculed him for sponsoring "the realism of a broken tea cup".) In 1881 he resigned from the *Atlantic* to devote his time exclusively to writing. He conducted the "Editor's Study" of *Harper's* Monthly from 1886 to 1891, and in 1900 took over the "Editor's Easy Chair" which he occupied until his death. His growing consciousness of social problems, his indignation at strike-breaking, and his study of Tolstoy, became manifest in *A Hazard of New Fortunes* (1890), which inserts Basil March between a capitalist and a socialist in a publishing

venture. *The Quality of Mercy* (1892) implies that capitalism is to blame for tempting an embezzler. *A Traveler from Altruria* (1894) finds America a disappointing contrast to his own Utopian country, in a novel where the narrative is an obvious vehicle for propaganda. *The Son of Royal Langbrith* (1904), is the story of a young man who worships the memory of his father, unaware that his father had cheated the family of the girl he loves. It successfully dramatizes the theme in one of Howell's best achievements with plot. He wrote a total of 37 novels, 2 volumes of short stories, 31 dramas ranging from blank-verse tragedy to farce, 11 travelogues, 4 volumes of verse, autobiographical essays, literary criticism, and a biography of Mark Twain. In his later years he was often considered the major living figure in American literature. He was a friend and adviser of both Twain and James, and he encouraged such younger writers as Stephen Crane and Hamlin Garland (qq.v.). His attitude toward Realism is presented in *Criticism and Fiction* (1891), where he proposes that reality in America, as opposed to the Naturalism of the French writers, must inevitably be a happy picture, for American life is, for the majority, a happy life. But he insisted that even the Realism of life's smile must be reinforced with moral lessons. It has been easy for many contemporary critics to ridicule the naivete of Howells; yet, in a few of his talents he surpassed many of our greatest writers, and he maintained a consistent integrity. He may be one of the next

19th-century writers to be "redis-
covered."

HUBBARD, Elbert (1856-1915), an Illinois salesman, was a failure in his literary
ambitions until he applied the promo-
tional techniques of salesmanship to editing the *Philistine* magazine (1895-1915),
platitudinously designed for Philistines,
at his artist colony near Buffalo, New
York. Here the Roycroft Press published
a series of 170 *Little Journeys* to the
homes of great men. He was the author
of *A Message to Garcia* (1899), a tale
of the Spanish-American War which was
incorporated in school books as a lesson
in duty. He perished in 1915, one of
the lesser casualties of the *Lusitania*.

HUGHES, Hatcher (1883-1945), won
the Pulitzer Prize for his *Hell-Bent fer
Heaven* (1924), a play of religious fana-
ticism among the Southern Appalachian
mountain people. *Ruint* (1925) is a
comedy dealing with the unconventional
predicament of a mountain girl. These
two plays were prominent in the folk-
drama interests of the 1920's and over-
shadow Hughes' other plays and collabor-
ations. Born in Cleveland County, North
Carolina, the dramatist graduated from
the University of North Carolina in 1907,
and from 1910 till his death was a
member of the English Department at
Columbia University.

HUGHES, [James] Langston (1902-),
spent his boyhood in Kansas City, Colorado Springs, Buffalo, Mexico, and Cleveland. He studied for a year at Columbia,
but returned to a varied career as an
international odd-job man. He taught
English and worked on his father's ranch
in Mexico; he farmed on Staten Island;
he was a doorman in Paris, a beachcomber in Italy, and a busboy in Washington, where Vachel Lindsay discovered
his poetry. Often using jazz rhythms and
a sardonic, colloquial expression, he
wrote of the problems of his race in *The
Weary Blues* (1926), *The Negro Mother*
(1931), *Shakespeare in Harlem,* (1942),
and several other volumes. He also wrote
a novel, *Not Without Laughter* (1930);
stories collected as *The Ways of White
Folks* (1934); 2 plays, *Mulatto* (1936),
and *Troubled Island* (1949); and an
autobiography, *The Big Sea* (1940).

Humanism, or "The New Humanism," is
a critical philosophy that was sponsored
by Irving Babbitt at Harvard and Paul
Elmer Moore (qq.v.) during the
1920's. A direct descendant of *Transcendentalism* (q. v.) in the history of American ideas, it is also a reaction against
the romanticism of the Transcendental
school, substituting a dualism of man
and nature for Emerson's monism. The
Humanist believes that human behavior
must be based on ethics, and that the
ultimate ethical principle is restraint, or
the Hellenic doctrine of reason leading to
a happy medium. Humanism is opposed
to scientific materialism with its emphasis
upon utilitarianism, and to sentimental
humanitarianism which relieves the individual of ethical responsibility. It sup-
ports Christianity as a means of achieving its ends. The term "humanism", is
also used to identify Renaissance ideo-

logy and contemporary political movements. (For further study, see CRITICISM; *Humanism*).

HUMPHREYS, David (1752-1818), a lieutenant-colonel and aide-de-camp to Washington, wrote pompous patriotic poems, and was associated with the "Hartford Wits."

HUMPHRIES, [George] Rolfe (1894-), was born in Philadelphia. He studied at both Stanford and Amherst. His many volumes of poems, written with irony and precision in conventional meters, include: *Europa and Other Poems and Sonnets* (1929), *Out of the Jewel* (1942), *The Summer Landscape* (1945), and *The Wind of Time* (1949). He has also translated *Poet in New York* (1940) and the *Gypsy Ballads* (1953) of Garcia Lorca. His translation of Virgil's *Aeneid* (1951) is considered by many Latin scholars to be the best in a long history of Virgil translations.

HUNEKER, James Gibbons (1857-1921), a music critic who also wrote general books on the arts, was associated with Mencken and Nathan. He also wrote a novel, *Painted Veils* (1920).

HUNT, Isaac (c. 1742-1809), born in Barbados, B. W. I., received his B. A. and M. A. degrees from the Academy of Philadelphia, and wrote pamphlets for the Loyalist cause before he was forced to flee to England. Leigh Hunt was his son.

HUNTER, Kermit (1910-), native West Virginian now living in North Carolina, is identified with outdoor historical pageantry, now known in the form he uses as "symphonic drama," because it employs ballet, singing, pantomime, as well as a definite plot. *Unto These Hills* (1949), given annually in the summers, tells the panoramic story of the Cherokee Indians; *Forever This Land* (1950) concerns Lincoln's New Salem days; *Horn in the West* (1952) is about pioneer migrations to Kentucky.

I

Imagism (c. 1908-c. 1925), an Anglo-American school of poetry, based on T. E. Hulme's modifications of Bergson's criticism. Ezra Pound (q. v.) joined Hulme and F. S. Flint in promoting the movement, and they were soon joined by Richard Aldington and his wife, H. D. (q. v.). Pound edited *Des Imagistes* (1914), an anthology which included, in addition to the leaders named, Amy Lowell and James Joyce. Miss Lowell became the leader of the movement in America, assisted by John Gould Fletcher and *Poetry* magazine. She published anthologies entitled, *Some Imagist Poets*, but never including Pound. Feeling that she had stolen his movement, Pound refined the creed further to "Vorticism". The Imagist seeks a hard, concrete analogy to immediate experience. His poem is not discursive, it is a dramatic, condensed metaphor, an "objective correlative" of an emotional experience. As a movement, Imagism gradually disappeared; but it purged poetry of a great deal of the 19th-century didacticism; it substituted the concrete picture for discourse; and its emphasis on free verse, although only a temporary influence, may have focussed a new interest on rhyme and meter when they returned to poetry. The Imagists were by no means the first or the only poets to break with the past or to employ aspects of their creed, but they assisted considerably in pioneering contemporary philosophies of poetry. (See also, CRITICISM).

Impressionism: was initially a school of French painting, represented by Monet, Pissaro, etc., who attempted to present the impression a scene makes on the eye rather than a photographic reality. In literature, for physical reasons, the definition can not be so precise, but those writers who emphasize moods and sensations rather than objective details have been called, or have called themselves, Impressionists. The literary movement was sponsored, again in France, by the Symbolists, Mallarmé, Baudelaire, and other imitators of Poe, who in turn influenced such Americans as Carl Sandburg, Amy Lowell, Marianne Moore, and Wallace Stevens. The prose of Huneker and Gertrude Stein, as well as Thomas Wolfe, has been classified as Impressionist. In the field of literary criticism, Impressionism is identified with the theories of J. E. Spingarn (q.v.).

INGRAHAM, Joseph Holt (1809-60), an Episcopal clergyman wrote, according to Longfellow, 20 novels in a single year. These ranged from such romances as *Lafitte: or The Pirate of the Gulf* (1836) to *The Pillar of Fire* (1859), a semi-Biblical story of Moses, interlarded with the author's personal observations.

INMAN, Henry (1837-99), wrote historical reports of the West, based on his adventures in Indian campaigns, which include *The Old Sante-Fe Trail* (1897) and *Buffalo Jones' Forty Years of Adventure* (1899).

IRVING, Washington (1783-1859), was the eleventh child of a wealthy New York merchant, and so much the "darling" of his family that he took his education at a series of mediocre private schools less seriously than his personal enjoyment of the books he selected to read in great number. After brief study in a law office, he began writing for newspapers edited by his older brother, and first attracted attention with his "Letters of Jonathan Oldstyle, Gent." (*Morning Chronicle*, 1802-3), a series of satirical sketches of New York society, which he had studied far more diligently than the law. Threatened with tuberculosis, he spent two years abroad (1804-6) in rest and casual study. On his return he was admitted to the bar, but practised intermittently while again satirizing New York in collaboration with his brothers and his brother-in-law, J. K. Paulding. Calling themselves the "Nine Worthies of Cockloft Hall," they produced a series entitled *Salmagundi; or, the Whim-Whams*

and Opinions of Launcelot Langstaff, Esq., and Others (1807-8). It has been suggested that when his fiancée, Matilda Hoffman, died at seventeen, he buried his grief and himself in *A History of New York, by Diedrich Knickerbocker* (1809), which satirized the Dutch, the Jeffersonian party, and literary pedantry in a single, coherent narrative. The charm and humor of the work argue against any notion that he was buried in grief, although the sorrow from his loss of the girl he loved, as well as from the loss of other girls who failed to love him, was surely perpetual. For the next several years he was occupied in business, politics, and miscellaneous editing. In 1815 his brothers sent him to Liverpool to manage a branch of their firm, but the failure of the business forced him to write to support himself. Scott encouraged him to write *The Sketch Book* (1819-20), which contained "Rip Van Winkle" and "The Legend of Sleepy Hollow," based on German folk tales, and informal essays concerning England, which he published under the pseudonym of Geoffrey Crayon, Gent. This work won him immediate fame, and he was lionized in London, Paris, and Dresden. His next work, *Tales of a Traveller* (1824), merited such adverse criticism that he was discouraged for two unhappy years, during which time he presumably was in love with Mary Godwin Shelley, while John Howard Payne, his John Smith, spoke for himself. The curious affair ended with his diplomatic assignment to Spain (1826-9), where he wrote his superb *Life of Columbus*, a popular but scholarly biography. He was next assigned to London as secretary of the United States Legation. Meanwhile, he published *A Chronicle of the Conquest of Granada* (1829), and *The Alhambra* (1832), a "Spanish Sketch Book". On his return to America in 1832 he was welcomed as the first American author to win international fame. A subsequent western tour provided materials for journals and *Astoria* (1836), concerning the fur trade. From 1842 to 1845 he served as minister to Spain, and after another year in London, he returned to his Tarrytown esate, "Sunnyside". The effort expended in writing his monumental *Life of Washington* (5 vols., 1855-59)

probably hastened his death, from heart attack, at 76. One of the greatest "minor writers" in our history, he is remembered as the "Father of American Literature," although his charm and sweet disposition make him an odd progenitor for much of the literature that followed him.

IRWIN, Wallace (1876-), a San Francisco journalist, wrote humorous verse collected as *Love Sonnets of a Hoodlum* (1902), *The Rubaiyat of Omar Kayyam, Jr.* (1902), etc.; the popular Hashimura Togo letters published as *Letters of a Japanese Schoolboy* (1909); and *Mr. Togo, Maid of All Work* (1913). He also wrote the novel, *Seed of the Sun* (1921).

J

JACKSON, Charles [Reginald] (1903-), worked in Chicago and New York as a book salesman until tuberculosis enforced a two-year vacation in Switzerland. His *The Lost Weekend* (1944) is a psychological study of a sensitive, charming alcoholic during five days of intense submission to the craving for drink. *Fall of Valor* (1946) portrays the destruction of a marriage by the homosexual tendencies of the husband.

JACKSON, Helen [Maria Fiske] Hunt (1831-85), was the daughter of a professor of Greek at Amherst. At the Ipswich Female Academy she made a lifelong friendship with Emily Dickinson. Her first publications were poetry: *Verses by H. H.* (1870) and *Sonnets and Lyrics* (1886). In 1852 she married Captain (later Major) Edward Bissell Hunt, with whom she lived on military posts until he was killed in 1863 while experimenting with his invention of a submarine gun. In Newport, she was encouraged by T. W. Higginson (q. v.) to continue her writing, and in the following years she produced poems, children's and travel books, and novels. In 1875 she married William Sharpless Jackson, a wealthy Colorado banker, and lived thereafter in Colorado Springs, except for research trips on behalf of her passionate indignation at the treatment of the Indians, reported in *A Century of Dishonor* (1881), and dramatized in her popular romance, *Ramona* (1884).

JAMES, Henry, Jr. (1843-1916), was born in New York City of Scotch-Irish ancestry, but received much of his early schooling abroad with his brother, William, who became the brilliant philosopher of Pragmatism. Henry James, Sr., a follower of Swedenborg in his theological writing, had designed his sons' education on a broad cultural and geographical basis. James attended the Harvard Law School, but he showed no interest in that profession, and he was encouraged by Howells to contribute reviews and stories to American periodicals. He made frequent sojourns to France and Italy, and in 1876 he settled permanently in London to devote himself to literature. His first novel, *Roderick Hudson* (1876), was already appearing serially in the *Atlantic Monthly,* as most of his future work was to appear, the first chapters reaching print before the final chapters were written. *The American* (1877) contrasts the American and French cultures, to the final disadvantage of the French. *The Europeans* (1878), a delightful light novel, brings expatriate Americans to New England, where the young man fits but his sister does not. *Daisy Miller* (1879) presents a charming American girl in Europe, while *Washington Square* (1881) presents a New York girl so unattractive that her father insists her suitor is after his money. In most of these early works, James was apparently obsessed with the distinction between the American and European cultures. A surface reading of his work, added to the fact of his choosing to live abroad, persuaded many of his contemporary critics that he was unduly critical of his homeland. Actually, he vacillated between praise and criticism. He was romantically in love with European culture and revolted by many aspects of the brash American scene. On the other hand, he recognized the superior virtues in the individual American, and respected his fresh honesty more than he condemned the naivete, while he was quite aware of the decadence of the aristocratic European. He persisted in the theme of Americans meeting Europeans at the most sophisticated levels of human intercourse, but with *The Portrait of a Lady* (1881), he reached a greater maturity of psychological realism and literary architecture in painting a young, conscientious American woman who rejects three varieties of good men for an exciting Machiavellian. *The Princess Casamassima* (1886), although his single "proletarian" novel, is perhaps singularly autobiographical in its deepest theme, the dispute between art and illogical duty. It is an autobio-

graphical equation that represents James and his family by characteristics rather than prototypes. The surface story concerns an illegitimate bookbinder meeting an illegitimate princess, once the heroine of *Roderick Hudson*, amidst an international anarchistic plot. *The Wings of the Dove* (1902) again brings an attractive American girl, this time a sick heiress, to Europe, and again she is the victim of the man she loves, but in a subtle symphony of motives arranged without a villain. *The Ambassadors* (1903) to France have the mission of rescuing a young New England gentleman from a Countess in what is perhaps the most perfect of the James novels. *The Golden Bowl* (1904) is a fugue in four voices: a father and daughter, the daughter and her husband, the father and his young wife, the daughter's husband and the father's wife. The four characters alternate their positions in the four groups in a counterpointed plot of desire and misunderstanding. As in many a James work, the reader is never certain just what has happened. Just before his death, James became a British subject in order to demonstrate his impatience with America for delaying her entrance into the First World War, which he regarded as a terrifying threat to his beloved civilization.

JARRELL, Randall (1914-) grew up in Nashville and took his A.B. (1935) and A.M. (1938) degrees at Vanderbilt. He has taught at Kenyon College (1937-9), Texas (1939-42), Sarah Lawrence (1946-7), Princeton (1951-2), Illinois (1953), and at the Seminar in American

SINCLAIR LEWIS

Photo: Erich Hartmann

LUDWIG LEWISOHN

JACK LONDON

AMY LOWELL

Civilization (1948) at Salzburg, Austria. He served with the Air Forces from 1942 to 1946. A critic, poet and novelist, his books include *Blood for a Stranger* (1942), *Little Friend, Little Friend* (1945), *Losses* (1948), *The Seven League Crutches* (1951), and *Poetry and the Age* (1953). *Pictures From an Institution* (1954) is a satirical novel concerning the faculty side of college life, featuring an unattractive lady novelist who teaches creative writing.

JEFFERS, [John] Robinson (1887-) was born in Pittsburgh, and spent part of his youth in Europe before his family moved to California when he was 16. He graduated from Occidental College, then investigated several occupations in graduate study, including medicine and forestry, until an inheritance permitted him to devote his life to poetry. He built himself a small castle overlooking the Pacific Ocean at Carmel, California, and gradually developed his unique attitudes toward poetry and society, independent of the movements which were beginning to dominate 20th century poetry. *Flagons and Apples* (1912) and *Californians* (1916), his first two volumes, presented conventional, romantic descriptions of California. In 1924, however, he brought out the first of a long series of narrative poems which would reveal a consistent social attitude and thematic pattern: *Tamar and other Poems* (1924), the other poems including "The Tower Beyond Tragedy," his personal version of the Electra legend. "Tamar" concerns California farmers, and introduced

through allegory his conviction that mankind suffers from an introversion of values and from self-deification. This allegory became persistent in his later work, which presented incest, rape, and murder leading to self-destruction. With a moral purpose, he stripped unmorality to its bare ugliness. The shorter poems in the first volume also established his lyric mood, his admiration for the enduring beauty of bare scenery and merciless nature. The allegory of incest and comparable violence, presented with Biblical or Attic overtones, is continued in *Roan Stallion* (1925), *The Woman at Point Sur* (1927), *Cawdor* (1928), *Dear Judas* (1929), *Thurso's Landing* (1932), *Give Your Heart to the Hawks* (1933), *Solstice* (1935), and *Such Counsels You Gave to Me* (1937), each volume containing the title narrative *and Other Poems. Descent to the Dead* (1931) contained elegies modeled on the Greek and again expressed his contempt for humanity. In 1938 *The Selected Poetry of Robinson Jeffers* was issued, followed by *Be Angry at the Sun* (1941) and *Medea* (1946). A few of his attitudes curiously parallel those of the more fashionable contemporary poets. He rejects humanitarianism, but extends this rejection much further than they in rejecting humanity itself. He deplores the modern emphasis on self-deification, but his remedy is pantheism rather than a moral absolutism. The pantheism and much of his poetic expression is romantic, yet his subject matter suggests the Greek classicism, even when the setting is California. He uses science merely to enforce his pantheism, for science suggests that man is of trivial importance in the great universe. In his prosody he has displayed great skill in experimenting with free verse, but in his exploitation of accentual prosody he is closer to the quantitative verse of the Greeks than to Whitman. His philosophy is Spenglerian rather than Thomistic. He shows great power, especially in his shorter poems. In his long narrative poems the characters are often so heavily burdened with allegorical implications that they lack dramatic motivation. The final appraisal of Jeffers may take more years than most of our contemporary poets will need, for he is the most brilliantly independent and startling of them all.

JEWETT, Sarah Orne (1849 - 1909), became well acquainted with Maine and its people as a girl when she accompanied her father, a distinguished physician, on his country visits. These experiences formed the material for her stories, collected in *Deephaven* (1877), and her novel, *A Country Doctor* (1884), concerning a woman physician who alternates between love and career. *The Country of the Pointed Firs* (1896), her masterpiece, studies the characters in a decaying seaport town with a quiet dignity. She was never married, and her life was as barren of adventure or passion as her work.

JOHNSON, Edward (1598-1672), was born at Canterbury, England, and emigrated to Boston in 1630. He wrote the first published history of New England, *The Wonder-Working Providence of Sions Saviour in New England* (1654),

which gives the spirit of the colony better than most chronicles. Puritan in its conception, the work includes considerable doggerel verse.

JOHNSON, Gerald W[hite] (1890-), historian and biographer, is also the author of two novels about the South: *By Reason of Strength* (1930), the chronicle of a staunch pioneering Scotswoman through several generations; and *Number Thirty-Six* (1933), the episodic story of a small town. His biographies are: *Andrew Jackson* (1927); *Randolph of Roanoke* (1929); *Roosevelt: Dictator or Democrat?* (1941); *Woodrow Wilson* (1944); *An Honorable Titan* (1944) on Adolph Ochs; and *The First Captain* (1947), about John Paul Jones. *The Secession of the Southern States* (1933), *America's Silver Age* (1939), *Our English Heritage* (1949), *Incredible Tale* (1950), *This American People* (1951), and similar books attempt, as do his other volumes, to trace basic American ideals through history and character. A native of North Carolina, Johnson went to Baltimore in 1926 where for seventeen years he was an editorial writer for the *Evening Sun*.

JOHNSON, Josephine [Winslow] (1910-), was reared on a Missouri farm, and was an artist and teacher before she won recognition as a novelist and short story writer. A realistic novel of farm life, *Now in November* (1934), was awarded the Pulitzer Prize, and her stories have appeared in the O. Henry and O'Brien collections. Her poems were published as *Year's End* (1937).

JOHNSON, Samuel (1696-1772), was the leader of the Church of England in New England, the first president of King's College (Columbia), and a disciple of the English philosopher, Berkeley. He wrote *Ethices Elementa* (1746) in a pleasing style, and satirized Calvin with paraphrases. Franklin published his work as *Elementa Philosophica* in 1752.

JOHNSTON, Mary (1870-1936), wrote historical romances and stories of her native Virginia, the best known being *To Have and To Hold* (1900), concerning an English noblewoman who weds a heroic Virginian after escaping a prearranged match with an English Lord, followed by adventures among pirates.

JONES, Howard Mumford (1892-), was born in Saginaw, Michigan, took his B. A. degree at the University of Wisconsin, (1914), and his M. A. at Chicago (1915). He received a D. Litt. degree from Harvard in 1936. He has taught English at Texas, North Carolina, and Harvard. His books include *A Little Book of Local Verse* (1916), *Gargoyles* (1918), *Ideas in America* (1944), and *Guide to American Literature* (1953). During World War II he served as special consultant to the program for reeducating German prisoners of war. His book of poems, *They Say the Forties* (1937), has been neglected, lost among the many volumes of Jones scholarship, but it deserves reprinting. Manipulating stock cliches with wit and perception, he produced a small collection of poems that reach beyond their surface of cynicism to nostalgia and even pathos.

JONES, Joseph Stevens (1809-77), began earning his living at ten, became an actor at eighteen, received an M. D. degree from Harvard at thirty-four, and wrote 200 melodramas, farces and comedies. His most popular play, *The Silver Spoon* (1852) exploited the shrewd Yankee farmer as a stock character.

K

KANTOR, MacKinlay (1904-), Iowa born journalist and novelist, has written detective stories for the pulp magazines, collected ballads (*Turkey in the Straw*, 1935), and contributed distinguished novels of the Civil War to our literature, including *Long Remember* (1934) and *Arouse and Beware* (1936).

KAUFMAN, George S. (1889-), has been, perhaps, the most prosperous playwright of the modern stage. He grew up in Pittsburgh, studied playwriting at Columbia, and worked at several routine jobs before becoming a New York journalist. In 1918 his first play was produced, and he has written more than twenty successful plays since then, collaborating with Marc Connelly, Ring Lardner, Edna Ferber, Morris Ryskind and George Gershwin, Katherine Dayton, and Moss Hart. *Of Thee I Sing* (1931) and *You Can't Take It With You* (1936) both won Pulitzer Prizes. Others of his highly successful plays are: *The Royal Family* (1927), *Dinner at Eight* (1932), and *The Man Who Came to Dinner* (1939).

KELLY, George [Edward] (1887-), actor, director, and playwright, won the Pulitzer Prize with *Craig's Wife* (1925).

KENNEDY, John Pendleton (1795-1870), a major politician and political figure in his day, practised law in Baltimore, served in Congress, and became President Fillmore's Secretary of the Navy. He wrote sketches, political satires, and novels which, though imitative of Cooper, displayed considerable narrative ability: *Horse-Shoe Robinson* (1835), and *Rob of the Bowl* (1838). His *Memoirs of the Life of William Wirt* (1842) reveals solid workmanship.

Kenyon Review (1939-), a quarterly literary journal edited by John Crowe Ransom and published by Kenyon College in Ohio. Most of its critical writers are representatives of the "New Criticism" (q.v.).

KEPHART, Horace (1862-1931), native Pennsylvanian and professional librarian, went to North Carolina after his health failed in 1904, "looking for a big primitive forest." For three years he lived in a lonely log cabin deep in the Great Smoky Mountains, and thereafter resided nearby in the village of Bryson City. He wrote numerous articles and books on the outdoors, adventure, history, and folklore. *Camping and Woodcraft* (1906) went through many editions. His principal work, however, was *Our Southern Highlanders* (1913), the classic on mountain whites.

KERR, Orpheus C., pseudonym of Robert Henry Newell (q. v.).

KEY, Francis Scott (1778-1843), a Maryland author, is remembered solely for *The Star-Spangled Banner*, which he wrote in 1814 after watching the bombardment of Fort McHenry from a British ship. His verses were set to the music of John Stafford Smith, an Englishman, and the song became the U. S. national anthem by act of Congress in 1931.

KILMER, [Alfred] Joyce (1886-1918), wrote *Summer of Love* (1911), *Trees and Other Poems* (1914), and *Main Street* (1917). He was killed in action during World War I.

KINGSLEY, Sidney (1906-), New York actor and playwright, attended Cornell where he took a conspicuous part in drama activities. His *Men in White* (1933) won the Pulitzer Prize. *Dead End* (1935) was a dramatic study of boys growing up in New York slums flanked by wealthy dwellings.

KITTREDGE, George Lyman (1860-1941), English professor at Harvard, and one of the great scholars in American letters, particularly of Shakespeare and Chaucer. In answer to the question of why he never took the Ph. D. degree, it is reported that he said, "Who would examine me?" It is indicative that this story is told to demonstrate his great knowledge, and never to suggest that he lacked modesty.

KNIGHT, Eric (1897-1943), was born in Yorkshire, England, and came to America at the age of 14. He was a journalist, scenarist, and trainer of jumping horses before he became a successful author. He wrote *Song on Your Bugles, This Above All,* and a novella, *The Flying Yorkshireman,* about a man who discovers that through faith he can fly like a bird. He wrote many stories about the Yorkshire miners, using their dialect for humorous enrichment, and he narrated many of them in dialect for radio programs. Serving as a Captain in World War II, he was killed in a plane crash in the jungles of British Guiana.

KNIGHT, Sarah Kemble (1666-1727), a Boston teacher, kept a journal of her five month trip to New York and return in 1704-5. This was published in 1825 by Theodore Dwight, and is frequently reprinted in reading anthologies of American Literature. The diary makes light of the hardships of the journey with a robust and pithy humor.

KOCH, Frederick Henry (1877-1944), born in Covington, Kentucky, was founder and director of the Dakota Playmakers at the University of North Dakota (1910-1918) and the Carolina Playmakers at Chapel Hill (1918-1944). His fame as a teacher of playwriting came from his origination and emphasis on the "folk-play," written about the simple, though not undramatic lives of the ordinary folk known to the student-playwright. Paul

Green, Thomas Wolfe, LeGette Blythe, Jonathan Daniels, Bernice Kelly Harris, and Frances Gray Patton are among the large number of established authors who emerged from his classes. Koch edited many volumes of one-act plays written by his students.

KOMROFF, Manuel (1890-), New York journalist, novelist, and short story writer, studied music, art, and engineering at Yale. During World War I he served as a foreign correspondent in Russia and the Orient. Many of his 130-odd short stories have been collected in *The Grace of Lambs* (1925), and *All in One Day* (1932). His novels include: *Coronet* (2 vols., 1929), a romance which chronicles the replacement of blooded aristocracy by the aristocracy of wealth from 1600 to 1919; a fictional account of the careers of the *Two Thieves* (1931) who were hanged with Jesus; a contrast of rich and poor in *New York Tempest* (1932); *The March of the Hundred* (1939) concerning lost men after World War I; a romance about *The Magic Bow*

(1940) of Paganini; and the *Echo of Evil* (1948) haunting a family with a murdered husband in its skeleton closet.

KREYMBORG, Alfred (1883-), a New York poet, playwright, novelist, and anthologist, is the author of *Plays for Poem-Mimes* (1918) and *Plays for Merry Andrews* (1920), both poetic dramas; and the novel, *I'm No Hero* (1933). *Funnybone Alley* (1927) was written for children. *Our Singing Strength* (1929) is a history of American poetry. His *Selected Poems* were published in 1945.

KRUTCH, Joseph Wood (1893-), one of the most distinguished of contemporary scholars, received his Ph. D. degree from Columbia in 1923, and taught there for many years. He was also drama critic of *The Nation*. His books include a study of *Edgar Allan Poe* (1926), an examination of *The Modern Temper* (1929), *Five Masters* (1930) of the novel, *The American Drama Since 1918* (1939), and a biography of *Samuel Johnson* (1944).

L

LA FARGE, Christopher, (1897-), grandson of the painter, John La Farge, is a painter, architect and poet. *Hoxie Sells His Acres* (1934) and *Each to the Other* (1939) are verse novels, and *Mesa Verde* (1945) is a verse drama about the cliff dwellers in southwestern Colorado. *The Sudden Guest* (1946) concerns a New England spinster's reaction to a hurricane. His war stories were collected in *East by Southwest* (1944).

LA FARGE, Oliver [Hazard Perry] (1901-), brother of Christopher La Farge, won the Pulitzer Prize with his novel about the Navajo Indians, *Laughing Boy* (1929). In addition to conducting archeological research in the southwest, and in Mexico and Guatemala, he has written several novels and a book of stories, most of them concerning Indian life. *The Enemy Gods* (1937) compares the Indian and white cultures.

LANGSTAFF, Lancelot, pseudonym of J. K. Paulding (q.v.).

LANIER, Sidney (1842-81), grew up in Macon, Georgia. His musical studies were interrupted by the Civil War, and 14 months as a prisoner of war left him consumptive. Impoverished at the war's end, he struggled to keep alive as a country lawyer. Knowing his life would be short, however, he finally determined to spend it with the arts. He became a flutist with the Peabody Orchestra of Baltimore, while writing poems and delivering lectures in literature. His *Poems* were published in 1877, and in 1879 he became a lecturer in English Literature at Johns Hopkins. *The Science of English Verse* (1880) and *The English Novel* (1883) were developed from his lectures. He held that the laws governing music are identical to those governing verse, a theory that accounts for both the beauty and the limitations of his poetry. The words combine to make beautiful music, but they say very little, and he was one of literature's worst offenders in the use of the *pathetic fallacy*. The poems he wrote to his wife have a beautiful tragic quality, and despite his use of conceits in his nature poems, he came very close to being a great poet. Certainly he was greater that the poets who were preferred during his lifetime.

LARDNER, Ring[gold] **W**[ilmer] (1885-1933), grew up in Niles, Michigan, and after studying engineering at the Armour Institute of Technology, became a journalist in Chicago and New York. He

began writing his "You know me Al"
sketches while conducting a sports column for the Chicago *Tribune,* and they
were published in 1916. Other books include *Bib Ballads* (1915), *Gullible's
Travels* (1917), *Treat 'Em Rough*
(1918), and *The Big Town* (1921). His
characters are normally ball players,
boxers, actresses, and others of the average American type, speaking the Mid-
west vernacular. His humor is flavored
with a cynicism concerning the proba-
bility of justice or logic in human affairs.

LAWSON, John (d. 1711), surveyor and
adventurer, published *A New Voyage to
Carolina* in London in 1709, although
it is now generally known by the inap-
propriate title, *Lawson's History of North
Carolina.* It contains the journal of Lawson's expedition into the hinterlands, a
description of North Carolina flora and
fauna, and an account of the customs
and manners of the Indians. Though obviously immigration propaganda, the
book possesses a sprightly style and is
one of the most readable of the early
travel narratives. The Scottish explorer
was put to death by the Indians on a
journey with the Swiss explorer Baron de
Graffenried, who escaped.

LAZARUS, Emma (1849-87), was educated by private tutors in surroundings of
wealth and culture, and began writing
poems in her early teens. Her work attracted the attention of Emerson, who invited her to Concord, and later led to a
friendship with Browning. Her early
work was characterized by a mature
melancholy, but after the Russian perse-

cution of the Jews she turned her energies as well as her writing to their defense. Of pure Spanish- Jewish stock, she
made a study of the Semitic race that led
to the composition of the verse drama,
"The Dance to Death," concerning 12th-
century Thuringian Jews, which is her
best work, but her other work also deserves to be better known, for her development of passion was deft and controlled. She was also an able translator
of Heine and of many medieval Jewish
writers. It is her sonnet that is carved on
the base of the Statue of Liberty.

LEARY, Lewis (1906-   ) took his B.S.
degree at the University of Vermont
(1928), and his M.A. (1932) and Ph.D.
(1941) degrees at Columbia, where he
is now a professor in the Graduate English department specializing in early
American literature. He has also taught
at the American University in Beirut, the
University of Miami (Fla.), and at Duke
University. His publications include
Idiomatic Mistakes in English (1932),
That Rascal Freneau (1941), *The Last
Poems of Philip Freneau* (1945) and
The Literary Career of Nathaniel Tucker
(1951). A specialist in American biblio-
graphy, he has also edited *Articles on
American Literature* appearing in period-
icals from 1920 to 1945 (1947). During
World War II he served with the Office
of Strategic Services in Washington and
Cairo.

LENNOX, Charlotte Ramsay (1720-
1804), although born in New York, was
sent to England at the age of 15, where
she lived the remainder of her life. Her

only novel of importance was *The Female Quixote; or, The Adventures of Arabella* (1752), which burlesqued the sentimental type of feminine novel which became her stock in trade. *The Life of Harriot Stuart* (1750), *The History of Henrietta* (1758), *Sophia* (1762), and *Euphemia* (1790) were all written in the pattern of her friend, Richardson.

LEONARD, William Ellery (1876-1944), for many years a professor at the University of Wisconsin, translated Lucretius and Beowulf, and wrote conventional *Sonnets and Poems* (1906), *The Vaunt of Man* (1912), *Two Lives* (1922) and an autobiography, *The Locomotive-God* (1927).

LEWIS, Alfred Henry (c. 1858-1914), was a Cleveland lawyer, a cow hand in the Southwest, a Kansas City journalist, and a Washington correspondent. He wrote six volumes of Wolfville stories under the pseudonym of Dan Quin, portraying the cattle and mining industries of the West in a humorous, colloquial style, including *Wolfville* (1897), *Wolfville Days* (1902), *Wolfville Nights* (1908), and *Faro Nell and Her Friends* (1913).

LEWIS, Janet (1899-   ), graduated from the University of Chicago in 1920, and married the critic, Yvor Winters, in 1926. She has written juveniles, short stories, three books of poems, and three novels: *The Invasion* (1932), *Against a Darkening Sky* (1943), and *The Trial of Sören Qvist* (1947).

LEWIS, Sarah Anna (1824-80), is remembered more as Poe's "Estelle" and benefactor than for her sentimental poems which were published as *Records of the Heart* (1844). She also wrote *Sappho* (1868), a poetic drama.

LEWIS, [Harry] Sinclair (1885-1951), was born in Sauk Center, Minnesota, the son of a physician. He interrupted his studies at Yale to work at Upton Sinclair's socialist colony and to seek work in Panama, but returned for his degree, and graduated in 1907. He engaged in editorial and newspaper work in Iowa, New York, San Francisco, and Washington. His first book was written for boys under the pseudonym, Tom Graham: *Hike and the Aeroplane* (1912). He became editor of an adventure magazine and wrote more than forty commercial short stories. His first serious novel, *Our Mr. Wrenn* (1914), was written in the H. G. Wells manner. *The Trail of the Hawk* (1915), portrays a barnstorming pilot, and captures the feeling that boys have for aviation, as *Free Air* (1919) captures their attraction to automobiles. In 1919 he borrowed $500 from his father and retired from all other activities to write his first masterpiece, *Main Street* (1920), satirizing middle-class life in the small town, as well as American middle-class ideologies in general. *Babbitt* (1922) satirizes American business practices and platitudes. With *Arrowsmith* (1925, Pulitzer Prize, 1926, but declined) he modified the intensity of his satire by presenting an idealistic scientist in opposition to medical "Babbitts,"

and the novel thereby gains in reality.
Elmer Gantry (1927) portrays a hypocritical clergyman, and *The Man Who
Knew Coolidge* (1928) ridicules the cigar-smoking type found in the railroad
smoking car. *Dodsworth* (1929) again
portrays an attractive protagonist, an
automobile manufacturer, who maintains
a practical idealism in the face of his
wife's infidelity. *Ann Vickers* (1933) is
a social worker, and *Gideon Planish*
(1943) is a too typical college dean.
It Can't Happen Here (1935) examines
America's weakness for Fascist types of
propaganda. *Cass Timberlane* (1945)
concerns the marriage of an established
judge to a young, immature girl. Lewis's
cynical satire of the American scene was
perhaps his greatest strength and weakness. It was so inclusive that it became
monotonous. However, when he abandoned satire for more human delineations of American life, his work lost
much of his power. He was the first
American (1930) to receive the Nobel
Prize for literature.

LEWISOHN, Ludwig (1882-), was
brought from Berlin to America as a
child, and grew up in Charleston, South
Carolina. He received his B.A., M.A.
and Doctor of Literature degrees from
the College of Charleston, and an M.A.,
from Columbia. He was a German professor at Ohio State (1911-1919) and
later an associate editor of *The Nation*.
His early novels (*The Broken Snare*,
1908; *Don Juan*, 1923; *The Case of Mr.
Crump*, 1926) concern problems of marriage, but his later work expresses his
belief that the Jewish race can make a

valid contribution to American culture
only through a faith in its own culture,
and not through assimilation (*The Island
Within*, 1928; *This People*, 1933). His
criticism (*Expression in Ameica*, 1932)
examines the unhealthful influence of
Puritan ideologies on American literature.

LINDSAY, [Nicholas] Vachel (1879-
1931), grew up in Springfield, Illinois,
and studied art in Chicago and New
York. In later years he lectured on art
and temperance and traveled as a "tramp
poet," exchanging verses for food. He
emphasized the oral quality of verse,
and became a popular public reader of
his own poems, dramatizing with gesture
and emphasizing his syncopated rhythms.
Of his many volumes of poetry, the best
known are *General Booth Enters into
Heaven and Other Poems* (1913), *The
Congo and Other Poems* (1914), *The
Chinese Nightingale and Other Poems*
(1917), and *Johnny Appleseed* (1928).
His friend, Edgar Lee Masters, wrote his
biography in 1935.

Local-Color, a term used principally in
America to describe fiction, and sometimes poetry, which portrays the regional
characteristics of a certain geographical
area, including living habits, the spoken
dialect, and the attitudes of the people.
It normally refers only to rural areas,
however, and too often is used to patronize good authors, or to justify a continuing interest in those who no longer deserve the attention they receive.

LOCKE, David Ross, see Petroleum
V. Nasby.

(130)

LOMAX, John A[very] (1872-1948), was one of the first serious scholars to collect American ballads, which he published as *Cowboy Songs and Other Frontier Ballads* (1910), *American Ballads and Folk Songs* (1934), *Our Singing Country* (1941), and *Negro Folk Songs as Sung by Lead Belly* (1936). He described his research work in *Adventures of a Ballad Hunter* (1947). He was assisted in making the later collections by his son, Alan Lomax (1915-   ).

LONDON, Jack [John Griffith London] (1876-1916), as a boy, scrambled for a living on the Oakland, California waterfront, as described in *Martin Eden* (1909). He bought a sloop and raided the oyster beds, later described in *The Cruise of the Dazzler* (1902) and *Tales of the Fish Patrol* (1905). At 17 he sailed on a sealing cruise which reached Japan. After tramping through the U. S. and Canada, studying briefly at the University of California, and examining the waterfront with a sociological interest, he joined the Klondike gold rush in 1897. Upon his return to Oakland he published stories of the Yukon in magazines, and they were collected as *The Son of the Wolf* (1900). Now both famous and popular, he produced a large number of stories and novels and became wealthy. He traveled extensively, often as a correspondent. His best novels were *The Call of the Wild* (1903) and *White Fang* (1906), both about dogs, but serving as metaphors concerning man and civilization; *The Sea Wolf* (1904), concerning a Nietzschean captain of a sealing ship exerting ruthless power over a gentleman and a lady; and his autobiographical novel, *Martin Eden*. He became interested in Socialism as a means of securing equality, but he was also curiously interested in both Marx and Nietzsche, and his admiration for both men may suggest, at least, that he foresaw the similarities of their theories as they have since been demonstrated by practical application.

LONG, John Luther (1861-1927), collaborated with Belasco in dramatizing *Madame Butterfly* (1897) which Puccini later operatized.

LONGFELLOW, Henry Wadsworth (1807-82), was born and reared in Portland, Maine, graduated from Bowdoin, and traveled for three years in Europe before becoming a professor at his alma mater. From 1836 to 1854 he taught at Harvard. His first wife died in 1835, and in 1843 he married Frances Appleton, whom he had met abroad, and they received a house as a wedding present from her wealthy father. She is the heroine of his poem, *Hyperion* (1839). In 1861 she was tragically burned to death when her dress caught fire. This was the major, and single tragedy in the poet's life. He achieved wealth, success, and incredible adulation practically without effort, for he produced his great body of work without strain, unimpeded by critical doubts. Perhaps his success stemmed from his ability to please a large and popular audience by mirroring their own attitudes and sentiments in skillful if uninspired verse. His sonnets are perhaps his best work, although he valued *Evangeline* (1847) above all his other poems. His translation of Dante's *Comedy* (1867) is valued now only for its literal quality. Yet he was honored by an L.L.D. from Cambridge and a D.C.L. from Oxford, and he is the only American honored with a placement of his bust in the Poet's Corner at Westminster Abbey. Ironically, his great popularity abroad created an early respect for American letters and an interest in American themes. His major publications were *Ballads and Other Poems* (1842), *Hiawatha* (1855), *The Courtship of Miles Standish* (1858), and *Tales of a Wayside Inn* (1863).

LONGSTREET, Augustus Baldwin (1790-1870), Georgia lawyer, educator, journalist and humorist, was educated at Yale, and served as a legislator and as judge of the superior court. Later he served as president of Emory College, Centenary College, the University of Mississippi, and the University of South Carolina. As a circuit lawyer in his youth he observed the manners, wit, customs, and dialect of the Georgia "crackers," and collected their "tall tales." These were the sources of his *Georgia Scenes, Characters, and Incidents* (1835), which first appeared in local newspapers under the pseudonym of "Timothy Crabshaw." Although the humor is Georgian, his techniques were in the tradition of Addison, and his work was a forerunner of later folk humorists, including Mark Twain. He also wrote a semi-autobiographical novel, *Master William Mitten* (1864).

LOTHROP, Harriet Mulford Stone (1844-1924), wrote children's books in the didactic tradition, but in a simple style that placed them above the average. Her *Five Little Peppers and How They Grew* (1881) sold more than 2,000,000 copies over a period of fifty years.

LOVETT, Robert Morss (1870-), professor of English at the University of Chicago, editor of *The Dial* (1919), and associate editor of *The New Republic*. With William Vaughan Moody he wrote a *History of English Literature* (1902). His critical studies include *Edith Wharton* (1925), and *A Preface to Fiction* (1930). He also wrote two novels, *Richard Gresham* (1904) and *A Winged Victory* (1907), and a play, *Cowards* (1914).

LOVINGOOD, Sut, see George Washington Harris.

LOWELL, Amy [Lawrence] (1874-1925), a sister of Abbott Lawrence Lowell (president of Harvard, 1909-33), traveled extensively from her Brookline, Massachusetts home from childhood on, to Europe, California, Egypt, Greece, and Turkey. Her first book of poems, *A Dome of Many-Colored Glass* (1912) was fairly conventional, but in 1913 she became acquainted with Pound, Fletcher, H. D., and Aldington in London, and became an enthusiastic sponsor of their Imagist movement. During the following years she took over the leadership of the movement in America. Her following work was characterized by experimentation, both in Imagism and in a free verse "polyphonic prose": *Sword Blades and Poppy Seed* (1914), *Men, Women, and Ghosts* (1916), *Can Grande's Castle* (1918), *Pictures of the Floating World* (1919), and *What's O'clock* (Pulitzer Prize, 1926). She had the organizing genius of a business man, as well as his habit of smoking cigars. Her *Critical Fable* (1922) was a witty and sometimes malicious *Who's Who* of contemporary poets. The strain of work which she put into her biographical study of *John Keats* (1925), added to her years of poor health, probably brought on her death in 1925 from a paralytic stroke.

LOWELL, James Russell (1819-91), the son of a Boston Unitarian minister, was educated at Harvard, made a poor start in the law profession, was unsuccessful in an attempt to establish a magazine, and contemplated suicide after an unrequited love affair, but his marriage in 1844 to Maria White changed his fortune for the better. She encouraged him in his writing of *Poems* (1844), *Poems, Second Series* (1848), *The Biglow Papers* (1848), *A Fable for Critics* (1848), and *The Vision of Sir Launfal* (1848); works which won him a reputation as poet, critic, and political satirist. Two years after the death of his wife (1853), he became a professor at Harvard and turned more to scholarship and criticism. In addition to teaching, he served as editor of the *Atlantic Monthly* (1857-62), and as an associate editor of *The North American Review* (1864-72). In

1877 he was appointed minister to Spain, and was transferred to the same position in England by Garfield (1880-85), where his natural conservatism and cultural attributes made him an outstanding success in diplomacy. His first wife had temporarily modified that conservatism, and he is remembered today chiefly for the work he wrote under her influence, before he reached the age of thirty. In his later years he was more important as a conservative critical force in American literature than as a creative artist. (See CRITICISM).

Lowell, Robert (1917-   ), great grandson of James Russell Lowell, grew up in Boston, and very early came under the influence of John Crowe Ransom and Robert Penn Warren. For a time he was married to the distinguished novelist, Jean Stafford. His highly allegorical poems, marked by a great technical skill, are too obscure to find a popular audience, but he received critical recognition with the award of the Pulitzer Prize for *Lord Weary's Castle* in 1947.

LOWES, John Livingston (1867-1945), Harvard professor (1918-1945), was a major American scholar, combining vital and imaginative perceptions with solid research and knowledge. He wrote *Convention and Revolt in Poetry* (1919); *The Road to Xanadu* (1927), an analysis of Coleridge; and *The Art of Geoffrey Chaucer* (1931), which remains one of the classic texts in Chaucerian scholarship.

LUHAN, Mabel Dodge (1879-   ), established salons in Italy (1902-12), New York City (1912-18) and in Taos, New Mexico, to which famous artists and writers have been continually attracted. In 1923 she married Antonio Luhan, her fourth husband, a Taos Indian, and she has found in the Indian race a refreshing contrast to the civilization of high society which she knew well in her earlier years. *Lorenzo in Taos* (1932) describes her intimate friendship with D. H. Lawrence. She has also published four volumes of her autobiography, *Intimate Memories.*

M

MacARTHUR, Charles (1895-), journalist, scenarist, and playwright, wrote *The Front Page* (1928), *20th Century*, (1932), and *Ladies and Gentlemen* (1939) with Ben Hecht. He is married to the actress, Helen Hayes.

McCULLERS Carson [Smith] (1917-), grew up in Columbus, Georgia, and seemed destined for a musical career when she went to New York at eighteen to study at Columbia and the Julliard School of Music. In 1937 she married Reeves McCullers, and in 1940, with the publication of *The Heart is a Lonely Hunter*, a study of a deaf mute, she won critical acclaim. *Reflections in a Golden Eye* (1941) is an incredibly perfect novel, concerning the psychological implications of murder in a peacetime army camp. *The Member of the Wedding* (1946), a tender study of the mind of a twelve-year old girl, was produced as a play in 1950. Her stories collected in *The Ballad of the Sad Cafe*, have frequently appeared in prize collections. Recently she has been active as a motion picture scenarist.

McCUTCHEON, George Barr (1866-1928), wrote more than forty popular novels, the best known being his *Grau-* *stark* series set in a mythical European court, and *Brewster's Millions* (1902) concerning a man who was required to spend a million dollars in one year, and encountered surprising difficulties.

McFEE, William [Morley Punshon] (1881-), an English ship's engineer, was the son of a British shipbuilder. He was educated in London, and after sailing on tramp steamers superintended by his uncle, entered the American merchant marine in 1911. After serving with the British Navy in the First World War, he settled at Westport, Connecticut to devote himself to writing. *Casuals of the Sea*, which publishers refused in 1912, was published in 1916. His other novels, all with sea or seaport environment, include *Captain Macedoine's Daughter* (1920), *Pilgrims of Adversity* (1928), *North of Suez* (1930), and *Ship to Shore* (1944). He wrote many other novels, as well as travel sketches and a book of verse.

McHUGH, Vincent (1904-), was born in Providence, Rhode Island, and became an excellent sailor on the surrounding waters. He went to New York as a young man, where he served on the staffs of the *Post* and *The New Yorker*. He served

for two years as editor-in-chief of the
Federal Writers' Project in New York
City, and later he was on the Writers'
War Board. During the war he also
served with OWI and the War Shipping
Administration. He has taught at New
York University and the University of
Denver, and he has been, by contrast, a
Hollywood scenarist. He has a unique
method of mixing fantasy with realism
in his novels: *Caleb Catlum's America*
(1936) is a comic satire on American
history that takes the protagonist through
more than two hundred years, meeting
Ben Franklin as well as Huck Finn; *I
Am Thinking of My Darling* (1943),
a best-seller, is a dead pan presentation
of New York City in the grip of a mysterious fever which removes the victim's
restraints and sense of responsibility. His
other novels are *Touch Me Not, Sing
Before Breakfast,* and *The Victory*
(1947). His intelligent and delightful
poems, reflecting his great interest in jazz,
were collected as *The Blue Hen's Chic-
kens* (1947). He has also written an
extremely valuable *Primer of the Novel*
(1950). Having tired of the eastern seaboard, he is now sampling the West,
and after trying Hollywood, Albuquer-
que, and Denver, he is now living in
San Francisco.

MACKAYE, Percy [Wallace] (1875-),
son of Steele Mackaye, graduated
from Harvard, and became a poet
and playwright. *The Canterbury Pilgrims* (1903) concerns Chaucer and his
Prioress. His subject matter ranged from
the classics (*Sappho and Phaon,* 1907)
to tales of the Kentucky mountaineers,

including *Gobbler of God* (1928). His
total production amounted to about fifty
published volumes. He was greatly interested in bringing folk culture and conscious literature together through the
medium of the spoken word.

MACKAYE, [James Morrison] Steele
(1842-94), New York actor and writer
of sentimental melodramas, dramatized
Tourgée's *A Fool's Errand.*

MacLEISH, Archibald (1892-   ), the
son of an Illinois merchant, attended
Yale and the Harvard Law School, and
was a captain of field artillery in the
First World War. He practised law in
Boston for three years, but in 1923 he
left the legal profession for five years
of travel and writing, living mainly in
France. Although he returned to reside
in America in 1928, he continued his
intermittent travels. He followed Cortez's route through Mexico in 1929
while working on *Conquistador* (Pulitzer Prize, 1933). He toured Europe and
Japan as a correspondent. His early subjective poems collected in *Tower of Ivory*
(1917), *The Happy Marriage* (1924),
The Pot of Earth (1925), *Streets in the
Moon* (1926), and *The Hamlet of A.
MacLeish* (1928), reflect the influence
of Eliot on both his technique and his
attitudes concerning the hopeless contemporary scene. With the onset of the
depression he began to incorporate a
sociological message in such verse as
Frescoes for Mr. Rockefeller's City
(1933). A superb radio play, *The Fall
of the City* (1937), revealed his awareness of the threat of totalitarian dicta-

torship. His *Collected Works* (1952) won the Pulitzer Prize. He served as Librarian of Congress from 1939 to 1945.

McNEILL, John Charles (1874-1907), a writer of humorous Negro dialect verse, was born near Wagram, North Carolina. He attended Wake Forest College, then practised law in his home state. In 1904 he joined the staff of the *Charlotte* (N. C.) *Observer*, and in it appeared a column of his essays and poems which attracted a wide following. His *Songs Merry and Sad* (1906) won the first Patterson Cup, given to a resident of North Carolina for a book showing "the greatest excellence and the highest literary skill." Most of the dialect verse appeared in *Lyrics from Cotton Land* (1907), whose six editions attest its popularity.

MAILER, Norman (1923-), grew up in Brooklyn, studied engineering at Harvard, and served with the Army in the Philippines and in Japan. After the war he traveled abroad and studied at the Sorbonne in Paris. His novel, *The Naked and the Dead* (1948), concerns thirteen men during an invasion of a Japanese-held island.

MALTZ, Albert (1908-), graduated Phi Beta Kappa from Columbia College and studied drama at Yale with George Pierce Baker (q.v.). A play written with George Sklar in the class, *Merry-Go-Round*, was produced in 1932. An analysis of civic corruption, it was so embarrassing to the New York City administration that it was closed for a week, whereupon public clamor forced its reopening. *Peace on Earth* (with Sklar, 1933) is an anti-war play. *Black Pit* (1935) concerns the mining union. His story, "The Happiest Man on Earth," won the O. Henry Memorial award in 1938, and has been widely reprinted in anthologies. His first novel, *The Underground Stream* (1940), contrasts the principles of a personnel manager and a Communist worker in the Detroit automobile industry. *The Cross and the Arrow* (1944) is a psychological examination of a German's self-conversion from Nazi ideology during World War II. *The Journey of Simon McKeever* (1949) pictures a septuagenarian seeking and finding hope for the future. As a Hollywood scenarist Maltz collaborated on "Destination Tokyo," and wrote "Pride of the Marines." His work has consistently championed the underprivileged and has called attention to social evils; yet he retains greater hope and dramatizes his themes better than most sociological writers.

MANNERS, John Hartley (1870-1928), English actor and dramatist, came to America in 1902, and enjoyed a successful career, writing more than 30 plays. His best known work, *Peg o' My Heart* (1912), was written for his wife, Laurette Taylor.

MARCH, William (William Edward March Campbell) (1893-), an Alabama hero of World War I, began writing short stories of his military experiences in 1928, which were published as *Company K* (1933). He has also written two novels, *Come In at the Door* (1934) and *The Tallons* (1936).

MARKHAM, Edwin [Charles] (1852-1940), California teacher and poet, is remembered mainly for a single poem, the popular "The Man with the Hoe," suggested by Millet's painting. He published *The Man with the Hoe and Other Poems* (1899), and *Lincoln, and Other Poems* (1901). He is said to be the prototype for Presley in Frank Norris's *The Octopus.*

MARQUAND, J[ohn] P[hillips] (1893-), Massachusetts author, presently living in New York, is one of the most successful of contemporary novelists. *The Late George Apley* (1937) was awarded the Pulitzer Prize, and later was produced as a successful play. His other novels include *H. M. Pulham, Esq.* (1941), *So Little Time* (1943), and *B. F.'s Daughter* (1946).

MARQUIS, Don[ald Robert Perry] (1878-1937), grew up in Illinois, and left school at 15 to work at a variety of occupations before he became a journalist. He became famous as a humorist for his columns, "The Sun Dial" in the New York *Sun,* and "The Lantern" in the New York *Tribune.* His best known book is *archy and mehitable* (1927), concerning a cockroach who types his observations by jumping on the keys of a typewriter, but can not engage the shift key, so that punctuation is spelled out and there are no capital letters. With Christopher Morley, his friend and admirer, he wrote *Pandora Lifts the Lid* (1924). His novels include *The Cruise of the Jasper B* (1916), *Off the Arm* (1930), and *Sons of the Puritans* (1939).

His comedy, *The Old Soak* (1922), was his only successful drama. He also published poems, stories, humorous verse, and informal essays.

Masses, The (1911-) was established in New York by Piet Vlag as a weekly journal of social criticism with socialist leanings. When Max Eastman became the editor in 1912 he intensified this political policy. After the journal was suppressed by the government, it was re-established as *The Liberator* (1919), and was increasingly radical, becoming affiliated with the Communist party in 1922. After another suspension (1924-6) it was again revived as *The New Masses,* and operated as a journal for the promulgation of communist doctrine in sociological and aesthetic matters. Thus, the original liberalism of the journal was replaced by a reactionary editorial policy, and writers with liberal imagination became increasingly scarce among its governors, sponsors, and contributors.

MASTERS, Edgar Lee (1869-1950), grew up in several small Illinois towns, attended Knox College for a short time, studied law in his father's office, and eventually became a successful Chicago lawyer. Meanwhile he was deeply interested in literature, and was a member of the literary community that flourished in Chicago between 1905 and 1915. He became suddenly famous in 1915 with the publication of his popular *Spoon River Anthology,* suggested by the *Greek Anthology,* and presenting the honest monologues of characters buried in a small-town cemetery. None of his many other volumes of poetry achieved the stature or the popularity of this first volume. He also wrote novels, and biographical studies of Lindsay, Whitman, and Twain. After 1926 he lived in New York City.

MATHER, Cotton (1663-1728), eldest son of Increase Mather, entered Harvard at the age of 12, and became an assistant to his father at the Second Church in Boston in 1681. Later he became co-minister, and in 1723 minister. He wrote 470 books, including histories, biographies, verses, sermons, and treatises on theology, medicine, and science, and none of them retain any value except as studies of a psychopathological case or of the limits of Puritan theology. *The Wonders of the Invisible World* (1693), as an example of his work, gives his observations concerning the work of devils in the victims of the Salem Witchcraft trials. Yet he championed smallpox inoculation, and was the first American to become a Fellow of the Royal Society. Taking himself seriously as a member of the "elect," perhaps his weakness and confusion lay in his complete inability to question his own divinity.

MATHER, Increase (1639-1723), the son of Richard Mather, was educated at Harvard, and took his M.A. from Trinity College, Dublin. He was president of Harvard from 1685 to 1701, where he encouraged scientific study while maintaining strict theological orthodoxy. He wrote more than 130 books on the history, politics, and theology of New England. He was successful during a mission to London in having Governor Andros replaced by Sir William Phips, who was in accord with the Puritan theocracy and conducted the Salem Witchcraft trials; however, Mather opposed the extreme measures taken by the new governor.

MATHER, Richard (1596-1669), emigrated to New England in 1635, where he became a prominent preacher and leader. He assisted in the writing of the *Bay Psalm Book* (1640). His most important work was *A Platform of Church Discipline* (1649), the keystone to New England Congregationalism.

MATTHEWS, [James] Brander (1852-1929), grew up in New York City, graduated from Columbia as class poet in 1871, and studied law, but chose rather to devote himself to literature and scholarship. He contributed widely to the periodicals, and wrote several historical studies of the theater, including *The Development of the Drama* (1903), *Molière* (1910), *Shakspere as a Playwright* (1913) and *Principles of Playmaking* (1919). He also wrote short stories and three successful plays in collaboration. He was a professor of literature and dramatic literature at Columbia from 1892 to 1924.

MELVILLE, Herman (1819-91), was the son of a cultivated New York merchant of distinguished family who died bankrupt in 1832. His only happy childhood days were spent on the prosperous farm of his mother's (Gansevoort) family. He was forced to leave school at 15, was a bank clerk, farmer, and teacher, and at

17 shipped as cabin boy to Liverpool, perhaps to escape his mother whom he found imperious under the hardship of being destitute with eight children. On his return he again taught school, at Pittsfield and East Albany. In 1841 he joined the crew of the whaler *Acushnet*, and sailed for the south seas on a voyage that was to provide the background for most of his great novels. The cruelty aboard provoked him to jump ship with a companion at the Marquesas, where he was held in a friendly but firm captivity by the Polynesian cannibals, later described in *Typee* and *Mardi*. He escaped them by boarding an Australian whaler, and he escaped the whaler at Tahiti, which he later described in *Omoo*. He worked as a field hand there, and later shipped to Honolulu, where he clerked and kept books for a merchant. The appearance of the *Acushnet* and her captain with affidavits concerning deserters may have been the persuasion for his signing as a common seaman on the naval frigate, *United States*, which sailed leisurely to Boston. Melville recorded the heroism and the brutality he witnessed on this 14 month voyage in his novel, *White-Jacket*. With a background and knowledge far different and more important for his purpose than he could have secured at any university, he settled down to write: *Typee* (1846), *Omoo* (1847), *Mardi* (1849), *Redburn* (1849) and *White-Jacket* (1850). He became a great success and famous personally as the man who had lived among the cannibals, although his satire on the type of civilization the missionaries were

bringing to the south seas alienated many a bigoted reader. He moved in the most distinguished literary circles of New York, and he journeyed to London and Paris. In 1847 he married Elizabeth Shaw, daughter of the chief justice of Massachusetts. In 1850 he moved to "Arrowhead," a farm near Pittsfield, Massachusetts, where he became a great friend of Hawthorne, who lived nearby, and where he wrote his masterpiece, *Moby Dick* (1851). His earlier, popular work had revealed both his narrative abilities and his penetrating mind, but with this Promethean novel he took readers further into the pursuit of eternal truth than they were willing or able to follow. There was, to be sure, the adventure story of Captain Ahab's frenzied pursuit of a great white whale, but there was also too much symbolism and ulterior meaning to be shrugged off. Readers who were accustomed to finding truth in such a neat package as "we can make our lives sublime" must have been nervous if not infuriated to find apprehensions of truth beyond the power of words to analyze. The book was not popular, and its reviews were even stupid. His next novel, *Pierre* (1852), which outdid Hawthorne in the psychological study of guilt, only helped to alienate his public. A year later the plates and the unsold copies of his books were destroyed by a fire at Harper's, and that was the virtual end of his publishing career. The whale had destroyed more than Captain Ahab. He retreated increasingly into a melancholy mysticism. He failed in attempts to get a consular position, and his efforts to support his family by lecturing were fruitless. From 1866 to 1886 he worked as a customs inspector, forgotten by public and friends. After publication of *The Piazza Tales* (1856), which contained the superb "Benito Cereno," and *The Confidence Man* satirizing commercialism (1857), he abandoned prose for poetry. In his last years he wrote one more story of good and evil at sea, *Billy Budd* (1924), recently produced as a play that charmed the critics but not the public. After Melville's death, *Moby Dick* was stripped to its bare narrative and published as a wholesome adventure story for boys. A hundred years after his birth date, a few critics, with Henry James and D. H. Lawrence behind them and Joyce and Virginia Woolf beside them, began to see and to write about his great accomplishment. And now, a hundred years after its publication, *Moby Dick* is the great novel that America has contributed to the literature of prophesy, taking its place beside the masterpieces of the world.

MENCKEN, H[enry] L[ouis] (1880-), Baltimore journalist, editor, and critic, became literary editor of the *Smart Set* (q.v.) in 1908, and co-editor with George Jean Nathan in 1914. In 1924 the two editors founded *The American Mercury* (q.v.), which Mencken edited for the next ten years. They announced that they would welcome manuscripts rejected by other publishers, and they were among the first to recognize the genius of Sherwood Anderson, Eugene O'Neill, Carl Sandburg, and Willa

Cather. Mencken has been the skeptic of everything in American life, but he praises our ability to burlesque pretentious dogmas. He has been a champion of truly American literature and has fought foreign influences. His best known work is *The American Language* (1919, revised in '21, '23, and '36; supplemented in '45 and '48), which mixes good if disorganized scholarship with diatribes against the English. His other books include: *George Bernard Shaw— His Plays* (1905), *The Philosophy of Friedrich Nietzsche* (1908), *A Book of Burlesques* (1916), *In Defense of Women* (1918), *Treatises on the Gods* (1930), and *Heathen Days* (1943).

MENKEN, Adah Isaacs (1835-68), a beautiful and captivating though incompetent actress, remembered more for her literary friendships—with Whitman and O'Brien in New York; with Twain, Harte, and Ward in San Francisco; with Dickens, Rossetti, Reade, and Swinburne in London; and with Gautier and Dumas *père* in Paris—than for her posthumously published poems, *Infelicia* (1868).

MILLAY, Edna St. Vincent (1892-1952), wrote poems for *St. Nicholas* while she was a Maine schoolgirl, and after her graduation from Vassar (1917), she published *Renascence and Other Poems,* followed by *A Few Figs from Thistles* (1920) and *Second April* (1921). *The Harp Weaver and Other Poems* was awarded the Pulitzer Prize in 1923. While living in Greenwich Village she supported herself by writing stories under the pseudonym of "Nancy Boyd." She was also associated with the Provincetown Players and the Theatre Guild, as an actress and playwright. In 1925 the Metropolitan Opera Association commissioned her to write the book for the Deems Taylor opera, *The King's Henchman* (1927). Later books include *Wine from These Grapes* (1934), and *Conversation at Midnight* (1937). Her work is characterized by great technical skill in conventional forms, treating subject matter from an early adolescent ecstasy in the joy of life to later disillusionment and bitterness. Whatever the subject, she has been one of the most popular poets of her generation.

MILLER, Arthur (1915-   ), grew up in Brooklyn, graduated from the University of Michigan (1938), and has combined his writing since then with periodic jobs to keep a sense of reality. His novel, *Focus* (1945), deals with anti-Semitism in a large corporation. *The Man Who Had All the Luck* (1944), his first Broadway play, was an apprentice work for the three powerful plays which he later produced. *All My Sons* (1947) won the Drama Critic's award. He won both the Drama Critic's award and the Pulitzer Prize for his sympathetic portrait of *The Death of a Salesman* (1949), which symbolizes the conflict between the American dream of "getting ahead" with the competitive reality. *The Crucible* (1953) dramatizes the Salem witchcraft trials.

MILLER, Henry (1891-   ), born in New York and now a resident of California, lived abroad, mainly in Paris from 1930 to 1940. His semi-autobiographical expressions of anarchy have been too frank in their physical descriptions to gain even the small audience that is ever loyal to more conventional cynics, but he does have a group of highly sophisticated admirers. His books include: *Tropic of Cancer* (1934), *The Cosmological Eye* (1939), *Tropic of Capricorn* (1939) and *The Air-Conditioned Nightmare* (1945).

MILLER, Joaquin, pseudonym of Cincinnatus Hiner Miller (1841-1913), arrived with his family in Oregon in 1852, and ran away three years later to work as a cook in the California mines. By 1857 he was living with the Digger Indians, had married a squaw, and had joined them in stealing horses. When he was captured, a friend sawed the bars of his cell, and he escaped back to his Eugene, Oregon home, where he attended Columbia College. In 1860 he was admitted to the bar, but instead of practising law he tried mining in Idaho, established a pony express, bought a newspaper and became a county judge. He married an Oregon poetess, "Minnie Myrtle," but she left him in 1867, and he left Oregon in 1870 to join the literary circles of San Francisco. The next year he sailed for England, where his western pose fascinated society, and with the publication of his two books of poems, *Pacific Poems* (1870) and *Songs of the Sierras* (1871), he was acclaimed the "Byron of Oregon." In the following years he traveled widely throughout the world, frequently as a correspondent, but maintained a home near Oakland. In 1883 he married Abbie Leland without bothering to divorce either of his first two wives. He wrote a play and several novels, and an autobiographical *Life amongst the Modocs* (1873). Today very little of his writing is considered important, but as a personification of the literary west living up to the expectations of the east and London, he remains a colorful figure in our literary history. Most of the tales he wrote or told about himself, however, were either exaggerations or prevarications.

MITCHELL, Donald Grant (1822-1908), also known as Ike Marvel, was a gentle New England editor and essayist who won fame with the publication of *Reveries of a Bachelor* (1850) and *Dream Life* (1851). In 1853 he married and subsequently had 12 children.

MITCHELL, Margaret (1906?-49), a Georgia newspaperwoman, spent ten years writing *Gone With the Wind* (1936), a romantic novel of the Civil War which was awarded the Pulitzer Prize and became the fastest selling novel in the history of American publishing. It sold 50,000 copies in one day, and a million and a half in the first year. Miss Mitchell avoided the literary limelight, however, and lived quietly in Atlanta with her husband, John R. Marsh. She was killed tragically in 1949 when an automobile struck her on an Atlanta street.

MITCHELL, S[ilas] Weir (1829-1914), Philadelphia surgeon, professor, writer in the field of medicine, and a novelist and poet, studied in Paris with the great physiologist, Claude Bernard, whose writing Zola later used as a basis for

Naturalism (q.v.). His poems were not exceptional, but his prose was divided between excellent historical novels and fine psychological character studies. His story, "The Case of George Dedlow" (*Atlantic Monthly*, July, 1866) was perhaps the first psychological study of a man in battle to appear in fiction. He used this theme also in his first novels, *In War Time* (1885), and *Roland Blake* (1886). His historical novels include *Hugh Wynne, Free Quaker* (1897), set in the American Revolutionary War; and a story of the French Revolution, *The Adventures of François* (1898). His other psychological novels include *Circumstance* (1901), about an adventuress; and *John Sherwood, Iron Master* (1911).

MONROE, Harriet (1860-1936), a minor Chicago poet and verse-dramatist, became known for her championship of original poetry as editor of *Poetry: A Magazine of Verse,* founded in 1912 (q.v.).

MOODY, William Vaughn (1869-1910), Indiana poet and playwright, achieved enviable success, after his graduation from Harvard, (1893) as a professor at the University of Chicago (1895-9 and 1901-7), as the author of *Poems* (1901), and as a dramatist. *The Masque of Judgment* (1900), and *The Fire Bringer* (1904), were the first two verse plays of an incomplete trilogy concerning man's justified rebellion against God, but neither was produced. His greatest commercial success was achieved with a contrast between frontier and Puritan traditions, *The Great Divide* (1909). *The Faith Healer* (1909) was too idealistic to achieve the same popularity. Moody's work was characterized by a spiritual idealism modified by recognition of man's conflicts in desire. He died before his less successful and pitied friend, Edwin Arlington Robinson, completely oveshadowed him in fame.

MOORE, Clement Clarke (1779-1863), Biblical scholar and author of *Poems* (1844), is remembered, if at all, for one poem, published anonymously in the Troy, New York *Sentinel* (Dec. 23, 1823); "A Visit from St. Nicholas," more commonly known by its first line, " 'Twas the night before Christmas."

MOORE, Julia A. (1847-1920), "The Sweet Singer of Michigan," wrote *The Sweet Singer of Michigan Salutes the Public* (1876), which attracted a reputation and won the praises of Mark Twain because her work is hilariously bad. Ogden Nash used her mistakes in versification as the basis for his comic style.

MOORE, Marianne [Craig] (1887-), was born in St. Louis, spent much of her childhood in Pennsylvania, and after graduating from Bryn Mawr (1909), and traveling in Europe, established her residence in New York City. She was editor of *The Dial* from 1925 to 1929. Her first poems appeared in *The Egoist*, the London magazine devoted to experimental work, and The Egoist Press published her *Poems* in 1921. Her *Selected Poems* were published in 1935, followed by *The Pangolin, and Other Verse* (1936); *What Are Years?* (1941);

and *Nevertheless* (1944). Her experimental prosody, based on a syllabic line, often using an obscure rhythm, is unique even in an epoch of experimentation. Her seemingly distorted metaphors give her work a freshness and a sense of wit. The content of her work is less original than her verse forms as it applies a moral philosophy to questions of contemporary behaviour.

MOORE, Merrill (1903-), was a member of the group of Tennessee poets that published *The Fugitive* (1922-5), and is presently a Boston psychiatrist (M.D., Vanderbilt, 1928). His prodigious production of sonnets has attracted more attention than the poems themselves, especially after the publication of *M* (1938), containing an even thousand sonnets. Earlier books include *It is a Good Deal Later than You Think* (1934) and *Six Sides to a Man* (1935).

MORE, Paul Elmer (1864-1937), taught Sanskrit and the classics at Harvard, edited *The Nation* (1909-14), and lectured at Princeton. He was a leader in the New Humanist movement, which he expounded in *The Demon of The Absolute* (1928). (See CRITICISM, *Humanism*, for his critical philosophy).

MORLEY, Christopher [Darlington] (1890-), New York journalist, essayist, and novelist, was a Rhodes Scholar at Oxford, the scene of his delightful little novel, *Kathleen* (1920). He has written more than 50 books, all of them marked by the author's very human personality, humor, and whimsy. His many collections of informal essays include

Mince Pie (1919), *Pipefuls* (1920), and *Forty-four Essays* (1925). His novels include two fantasies, *Where the Blue Begins* (1922) and *Thunder on the Left* (1925), as well as his more realistic novels, *Human Being* (1932) and *Kitty Foyle* (1939). Morley never tackles world-shaking problems, but his great love of the human race, his charm, and the quality of revelation he achieves in dealing with ordinary experience, combine to give his work a quality of greatness that is reminiscent in a small way of Chaucer.

MORTON, Sarah Wentworth (1759-1846), under the pseudonym of "Philenia, a Lady of Boston," wrote *Ouâbi: or, The Virtues of Nature* (1790), and *Beacon Hill* (1797) in heroic couplets. She was thought to be the author of *The Power of Sympathy* (1789), often named the first American novel, until late 19th-century scholars determined that it was written by William Hill Brown (q.v.).

MUMFORD, Lewis (1895-), a New York critic of the arts and man in society, wrote *The Story of Utopias* (1922), *Herman Melville* (1929) and *The Golden Day* (1926), in addition to several sociological studies.

MUNSON, Gorham B. (1896-), New York author, editor, and critic, has written studies of *Waldo Frank* (1923), *Robert Frost* (1927), *Destinations: A Canvas of American Literature Since 1900* (1928), *Style and Form in American Prose* (1929), and *The Dilemma of the Liberated* (1930).

MURFREE, Mary Noailles (1850-1922), became a partial cripple as a result of an illness when she was a girl in Murfreesboro, Tennessee, a misfortune that caused her to turn to study and later to writing. She became the first to write about the southern mountaineers, whom she described so accurately and whose dialect she reproduced so faithfully that her work has a quality of realism. She contributed stories to the magazines under the pseudonym of Charles Egbert Craddock for 6 years before her first collection was published, *In the Tennessee Mountains* (1884), and both her editors and the public assumed she was a man. She published more than a dozen additional collections of her dialect stories, and several historical novels.

N

NASH, Ogden (1902-), New York author of comic verse, often printed in *The New Yorker,* and collected into several volumes including *Free Wheeling* (1931), *The Bad Parent's Garden of Verse* (1936), *I'm a Stranger Here Myself* (1938), and *The Face is Familiar* (1940). Nash acknowledges his debt to Julia Moore (q.v.), whose incredibly bad prosody became his humorous device. His major contribution to humorous verse is the variable line: he simply keeps the line going until he is able to arrive at the rhyme he needs. Inverted platitudes often provide his argument, or he may engage in a relative analysis of the advantages of bathing before shaving as opposed to shaving before bathing. He directs his work toward a more sophisticated audience than do most humorous versifiers.

NASBY, Petroleum V., pseudonym of David Ross Locke (1833-88), Civil War newspaper humorist, who used the stock devices of deformed grammar, punctuation, and logic in ridiculing the southern cause by pretending to support it. The letters which appeared in the Findlay, Ohio *Jeffersonian* were published as *The Nasby Papers* (1864).

NATHAN, George Jean (1882-), editor and dramatic critic, grew up in Indiana, and was educated at Cornell and the University of Bologna. With H. L. Mencken (q.v.) he edited the *Smart Set* (1914-23) and *The American Mercury* (1924-30), and wrote three plays. He is more interested in the beauty, the color of life than in its problems. His drama reviews frequently are more entertaining than the subject plays. In addition to numerous collections of his essays, he has written an *Encyclopaedia of the Theatre* (1940) and the annual *Theatre Book of the Year* (1943-). His epigrammatic satire includes *The Autobiography of an Attitude* (1925) and a book of advice for children to *Beware of Parents* (1943). An early champion of O'Neill and a constant foe of sentimentality or intellectual sham, he has exerted a healthful influence on the 20th-century theater.

NATHAN, Robert [Gruntal] (1894-), of a prominent New York family, was educated in the best American and European private schools, and at Harvard. He is an accomplished musician, and skilled in athletics. In writing his short novels he is a diligent craftsman as well as a witty satirist with a saving human sympathy. The best known of his numerous novels are: *The Bishop's Wife* (1928), concerning an angel in love with a bishop's wife (and appearing very realistic in a business suit in the motion picture version); *One More Spring* (1933); and *Portrait of Jenny* (1940), about a painter finding inspiration in a child. Mr. Nathan has also written formal poems which have been collected in five volumes.

Naturalism, as a critical term, was originally applied to that literature which examined its characters and their actions with scientific objectivity. Zola, considered to be the originator of Naturalism, declared that literature could derive an example from the medical experiments of Claude Bernard, that the novelist could set up an experiment within his novel by moving his characters to show that the succession of facts depicted would be those which the determinism of the phenomena under examination demanded (*The Experimental Novel*). The Naturalists believe that man is controlled by his passions, instincts, circumstances, and environment. In general usage, however, the term Naturalism has been applied to those works which depict man under the control of the most unhappy environment and the most base passions. The term is frequently used to classify those novels which carry a message in favor of social change for the improvement of man's economic circumstances. See also, CRITICISM, *Early Twentieth Century*.

NEAL, John (1793-1876), whose father died a month after his birth in Portland, Maine, was forced to work for a living from childhood. He was a merchant in Baltimore in his twenties, and was admitted to the Maryland bar in 1819. Meanwhile, he had edited the Baltimore *Telegraph* and compiled a history of the American Revolution. During this period he published several poems and miscellaneous books, and the novel, *Randolph* (1823). He spent 4 years in England, where he became the first American to contribute regularly to English periodicals, and served as secretary to Jeremy Bentham. Soon after his return, he settled in Portland, where he practised law, lectured, engaged in various businesses, and was an editor and writer. His works include *Seventy Six* (1823), *Brother Jonathan* (3 vols., 1825), *Rachel Dyer* (1828) and *The Down-Easters* (1833). Although he showed power and an original mind, he wrote too hastily to produce anything of value.

NEIHARDT, John G[neisenau] (1881-), grew up in Kansas, was a teacher and farmer, and worked with the Omaha Indians (1901-7). This experience, combined with historical research, provided the material for his western stories, essays, poems, and one novel. *The Song of Hugh Glass* (1915) is an epic poem about a frontier trapper; *The Song of Three Friends* (1919) concerns the Ashley-Henry exploratory expedition to the Green River; *The Song of the Indian Wars* (1925) recounts the passing of the buffalo; *The Song of the Messiah* (1935) refs to the Indian hope for a leader to rescue their tribes from annihilation. His *Collected Poems* (1926) contain his lyrics. His story collections include *The Lonesome Trail* (1907), and *Life's Lure* (1914) is a novel about mining in the Black Hills.

NEWBERRY, Julia Rosa, (1853-1876), the daughter of a wealthy Chicago capitalist, and a frequent traveler to Europe during her short life, may have become the great novelist she aspired to be had she not died in Rome of a throat infection at the age of 22. *Julia Newberry's Diary* (discovered in 1932, published in 1933) reveals a keen wit, a fine prose, and a realistic perception of the life she saw.

New Criticism, The, the name given to a critical movement sponsored by Allen Tate, Robert Penn Warren, Cleanth Brooks, John Crowe Ransom, and others, which has come to dominate the critical attitudes among academic circles in the past two decades. These critics deplore the loss of an ethical tradition and the corresponding loss of the moral and intellectual authority of the poet. The enemies of tradition are the new scientific materialism and romantic individualism. In matters of technique the *New Critics* favor what Tate calls "tension," a metaphysical combination of intension and extension to achieve complications of metaphor. See *New Critics* under CRITICISM for a fuller explanation of this philosophy.

New Directions (1936-   ), annual journal of experimental literature published by the New Directions press.

New Masses, see MASSES.

NEWELL, Robert Henry (1836-1901), wrote *The Orpheus C. Kerr Papers* (5 vols., 1862-71), satiric interpretations of public life during and after the Civil War, written in the cracker-box style of humor so prevalent in that period. He was one of the husbands of Adah Isaacs Menken (q.v.). Under his own name Newell wrote romantic fiction and sentimental verse.

New England Courant, The (1721-26), the lively journal founded by James Franklin, and published by his half-brother, Benjamin Franklin, while James spent a month in jail for his criticism of religious and civil leaders.

New England Magazine, The (1831-5), a precursor of the *Atlantic,* published Hawthorne, Longfellow, Holmes, and Whittier.

NEWTON, A[lfred] Edward (1863-1940), Philadelphia book collector who wrote charming essays concerning his avocation, collected in several volumes including *The Amenities of Book-Collecting and Kindred Affections* (1918), *This Book-Collecting Game* (1928), and *End Papers* (1933).

New Yorker, The (1925-   ), weekly magazine of sophisticated taste, subtle humor, and well-balanced investigation of the contemporary scene. It was founded by Harold Ross, who was publisher until his death in 1952. It has practically established many contemporary writers whose work was too special for more commercial magazines. Contributors include Robert M. Coates, Sally Benson, S. J. Perelman, Ogden Nash, Dorothy Parker, Jean Stafford, and John Hersey, whose objective report on *Hiroshima* occupied an entire issue. Its cartoons have served to advance in sophistication the American taste for humor, an influence that extends to readers who never see this magazine, but receive their conditioning second hand from other journals. Cartoonists who publish almost exclusively in *The New Yorker* have included Peter Arno, Helen Hokinson, Whitney Darrow, James Thurber, and Rea Irvin. E. B. White has been mainly responsible for the style of "The Talk of the Town," highly polished to a casual spontaneity. Collections of work appearing in the magazine have been published as *The New Yorker Album.*

NICHOLS, Thomas Low (1815-1901), New Yorker reforming journalist, wrote *Forty Years of American Life: 1821-1861* (1864), valuable as a source of detailed information concerning the life of that period. He also wrote three novels: *Ellen Ramsay* (1843), *The Lady in Black* (1844), and *Raffle for a Wife* (1845).

NICHOLSON, Meredith (1866-1947), Indiana novelist and biographer of James Whitcomb Riley in *The Poet* (1914), wrote numerous novels including *A Hoosier Chronicle* (1912), as well as stories, essays, and poems.

Nobel Prizes for literature have been awarded to the following American authors: Sinclair Lewis (1930), Eugene O'Neill (1936), Pearl Buck (1938), T. S. Eliot (1948), William Faulkner (1949), and Ernest Hemingway (1954). An international competition, the prizes were established to recognize significant contributions to science, literature, and peace through the bequest of Alfred B. Nobel (1833-96), a Swedish scientist. The prizes, awarded annually since 1901, consist of grants of about $40,000 to each winner.

NORDHOFF, Charles (1830-1901), was brought from Prussia to Chicago at the age of five, where his father was a pioneer fur-trader. After three years as a seaman, traveling around the world, Nordhoff became a New York journalist, and described his adventures in a series of three books collected as *Nine Years a Sailor* (1857). He also wrote non-partisan economic and political studies, including *Communistic Societies in the United States* (1875).

NORDHOFF, Charles Bernard (1887-1947), grandson of Charles Nordhoff (q.v.), resided in Tahiti, and wrote popular novels in collaboration with James N. Hall, including *The Hurricane* (1936), *Mutiny on the Bounty* (1932), and *Pitcairn Island* (1934), all as successful on the motion picture screen as they have been in magazine and book form.

NORRIS, Charles G[ilman] (1881-1945), brother of Frank Norris, married the popular novelist, Kathleen [Thompson] Norris (1880-   ). His novels investigate sociological problems, including feminism, birth control, and business ethics. They include *Salt* (1917), *Brass* (1921), *Bread* (1923), *Pig Iron* (1925) and *Flint* (1944). His novel *Bricks without Straw* (1938), may be distinguished from that of the same name by Albion W. Tourgée (q.v.) by the fact that "without" is not capitalized in the Norris title.

NORRIS, Frank (Benjamin Franklin Norris) (1870-1902), moved with his parents from Chicago to San Francisco in 1884. While studying art in Paris he wrote medieval romances, and while studying at the University of California (1890-4) he wrote *Yvernelle, A Tale of Feudal France* (1892). Under the influence of Zola, he turned abruptly to Naturalism (q. v.) after graduation, and while studying at Harvard wrote *McTeague* (1899), a proletarian novel about a brutal dentist. He reported the Boer War for *Collier's* and the San Francisco *Chronicle* from South Africa in 1895-6, and after a year at home he went to Cuba to cover the Spanish American War for *McClure's.* Meanwhile, he vacillated from Zola to Jack London's style in *Moran of the Lady Letty* (1898) and *A Man's Woman* (1900). He next planned an "Epic of Wheat", and visited a California wheat ranch to secure material for *The Octopus* (1901), which depicts the growing of the wheat against a background of strife between the ranchers and a railroad company. *The Pit* (1903) deals with speculation in wheat on the Chicago Board of Trade, with the speculator's romance and marriage problems for story. Norris

died suddenly after an appendix operation, and was unable to complete *The Wolf,* which was to show the consumption of the wheat in a European village stricken with famine. His collected *Works* (10 vols., 1928) contain other stories and articles previously unpublished in book form. Norris is considered to be one of the leading pioneers in the development of the realistic American novel.

North American Review, The (1815-1939), a Boston literary and scholarly journal, printed the work of Bryant, Emerson, Irving, Longfellow, Lowell, Holmes, Howells, Whitman, Twain, and Henry James. Among its editors were Lowell, Henry Adams, and H. C. Lodge.

NOVEL, THE AMERICAN:
   Revolutionary Period: *The Power of Sympathy* (1789) by William Hill Brown (q.v.) is generally agreed to represent the first American novel. The selection of any work for this distinction must be somewhat arbitrary, and it can only be said that Brown's novel best deserves the distinction. The American born author, Charlotte Lennox, had published *The Female Quixote* in 1752, but by then she had become entirely English with respect to both her residence and her career. This was true also of Edward Bancroft, who published *The History of Charles Wentworth Esq.* in London (1770). By contrast, Susanna Rowson, who was born in England, later became recognized as an American novelist, but only after the publication of *Charlotte Temple* (London, 1791;

New York, 1794), so that her three earlier English novels do not signify. The history of the American novel therefore begins about fifty years after the English novel had its inception with the works of Richardson, Fielding and Smollett. It was not by any means an auspicious beginning, for *The Power of Sympathy* was notably lacking in the power of narrative. Brown's involved rhetoric and moral didacticism are no less distasteful than his weaknesses in structural arrangement, characterization, motivation, sincerity and true morality. His announcement that the book was intended "to represent the specious causes, and to expose the fatal consequences, of seduction; to inspire the female mind with a principle of self complacency, and to promote the economy of human life," unfortunately was an announcement of the fiction pattern for the next three decades. If the American novel was launched fifty years after the English, it was starting from the same point as that from which the English had started, with the literary attitudes of Richardson.

America did not provide a fertile soil, in the 18th century, for any of the departments of literature, and the climate was especially poor for the development of the novel. While the colonies were political dependents of England, there was no national compulsion to develop a literature, for the settlers were English subjects, and they could take pride in the great tradition of English literature. After they became an independent nation, however, the British heritage was not the valuable foundation for a new literature that many critics have assumed. The new soil and the revolutionary ideologies needed a matching literary convention, but no new genius appeared to supply this need. Most early American literature was a weak imitation of the British, and there were many factors to discourage the development of the literary art. The immediate problems of the new country were economic and political, and genius was naturally diverted to these fields. Alexander Hamilton had early planned for a literary career, but circumstances directed his remarkable talents into making instead of imitating history. There was no international copyright law until 1891, and American printers could publish the most popular English novels without the expense of compensating the English authors. American authors therefore needed to pay for the printing of their own books. Nor were there any means of publicizing and selling the unknown books on a profitable scale. The South, which had the aristocratic environment that was more favorable to leisure, and therefore to art, than severe New England, also had the aristocratic antipathy to mass education, so that the reading market was restricted. In more literate New England the Puritan mores were a militant block against the production of any fiction, and the fiction which did get past the block was devitalized by the contest. Puritanical opposition to fiction was not confined to the Boston area, for the *Philadelphia Repository and Weekly Register* called novels "the one great engine in the hands of the fiends of darkness" (June 6,

1801). It must be remembered that even in England the novel was looked upon as a mere and sometimes degrading form of entertainment until well into the 19th century, not the sort of book that a serious and intellectual person would be apt to read. Every new art form suffers from prejudice, and a comparison may be made with 20th-century attitudes, first toward motion pictures, and now toward television. But to the Puritan patriarchs the novel was worse than a waste of time; it poisoned the minds of young girls, inflamed their passions, and led to seduction. The novel was also condemned in a general way as a threat to both religion and democracy, but any listing of the specific dangers almost always referred to young ladies. The results of these condemnations, principally from the pulpit, were paradoxical. The early novel became exactly what it was denounced for being, while advertising itself as a moral tract. The Calvinist doctrine of "the elect" was still sufficiently strong in Puritan minds to enforce their need for security in distinctions of good and evil. A person was either good or bad, and any delineation of character or psychological motivation of action could not be tolerated. The novelists simply made the most of the situation. In their prefaces they attacked all other fiction as evil, using the very language of the pulpit, but they assured readers that their work had been written for the very purpose of promoting morality, by acquainting young ladies with the evil intentions and satanic machinations of the seducer, and showing the fearful consequences of sub-

mission. Each novelist proceeded to lead his heroine through a series of adventures so interesting that the sale of his books was assured, while he won the plaudits of the clergy by larding the narrative with moral counsel and depositing the seduced heroine in a watery grave at the climax. In *The Power of Sympathy*, Brown's two lovers are barely prevented from incest by learning of the adultery of their mutual father, and their story is embellished by two incidental plots concerning a seduction and an abduction. Mrs. Rowson displayed an unusual tenderness for her erring heroine, *Charlotte Temple*, and wrote the best of the early novels, but Charlotte quite properly died in childbirth. So also did *The Coquette* (1797), in the novel written by Hannah Foster.

During this period the Gothic romance was having a great vogue in England, but the Gothic concept was alien to a democracy, and this type of novel was not a good vehicle for moral dogma, so that its vogue in America was limited. Mrs. Sally Wood flavored her novels with Gothic concepts, especially her *Julia and the Illuminated Baron* (1800). Insofar as America produced an indigenous Gothic novel, it is to be found in the works of Charles Brockden Brown (q.v.), and he produced the effects rather than the trappings of this style.

The sentimental novel as a whole was aptly parodied by Mrs. Tabitha Tenney in *Female Quixotism: Exhibited in the Romantic Opinions and Extravagant Adventures of Dorcasina Sheldon* (1801). Hugh Henry Brackenridge (q.v.) also used the Quixotic formula in satirizing

Photo: A. Aubrey Bodine

H. L. MENCKEN

EUGENE O'NEILL

EDWIN ARLINGTON ROBINSON

Photo: William A. Smith

CARL SANDBURG

American life with *Modern Chivalry,* but despite his fine prose style and penetrating perceptions of our early democracy in action, he stands as an accidental ornament without much relationship to the development of the novel. Indeed, it may even be questioned whether his episodic book may be called a novel, except in the sense that it is a long work of fictional episodes. Royall Tyler's *The Algernine Captive* (1797) was also picaresque in structure, satirical in tone, and far superior to the average novel of the day. If Brackenridge and Tyler do not fit into the chronological schematism of the American novel, they adumbrate the intellectual novels of the future, as Charles Brockden Brown foreshadows Poe and Bierce. It was their virtue that they were not representative of the times.

Nineteenth Century: James Fenimore Cooper stands as the first American novelist to create his fiction from the elements of the American scene, and with Irving he marks the first worthy contribution that his country made to international literature. Earlier writers had tried the American historical romance, but neither their popularity nor their literary achievment was sufficient to give them historical prominence. Although Cooper's literary quality is still debated, his originality and popularity can not be questioned. It is significant to this study that his defenders compare his novels to ancient epics, and that D. H. Lawrence advised us to read *The Deerslayer* "as a lovely myth." The novel of the frontier has continued to present a myth and to treat American history in

epic terms, whether in the popular tales of Zane Grey or the literary histories of H. L. Davis. Since Cooper carried forward many of the attitudes of the earlier sentimental romance, and set the pattern for the American myth of the future, he may be the most representative novelist in and of the development of this medium. He still overshadows his immediate contemporaries, many of whom followed his pattern of the historical romance: John Neal, Lydia Maria Child, James Kirke Paulding, and Catherine Maria Sedgwick. His most distinguished and original disciple was William Gilmore Simms, who was less sentimental and more representational in his portraits of both men and history. Simms' picture of the American Indian is far more accurate than Cooper's. Robert Montgomery Bird took the historical romance to a Mexican setting, portraying the adventures of Cortez, and Joseph Holt Ingraham treated *Lafitte, The Pirate of the Gulf* (1836).

If the American novel had reached a degree of competence when Cooper published *The Spy in* 1821, it was blossoming into maturity thirty years later as Melville settled in Pittsfield, Massachusetts, to write *Moby Dick,* and match philosophies with Hawthorne, six miles away in Lenox, who had just achieved success with *The Scarlet Letter.* Hawthorne had printed an earlier novel, *Fanshawe,* at his own expense in 1828, and the reception discouraged him from trying this form during the next two decades. Like Cooper's work, *Fanshawe* owed certain debts to Scott, but Hawthorne could hardly compete with the romancers, being one of the most limited if one of the greatest of American novelists. With *The Scarlet Letter* he found a story suitable to both his temperament and his talents, and once set in the right direction, he was able to write the first "perfect" American novel and to embody in it a mature philosophy concerning human behavior. Melville, who had written competent novels that carried the adventure story further than any of his readers suspected, drank brandy and talked "ontological heroics" with the older Hawthorne during the period in which he was struggling to shape his powerful story of the whale of good and evil. *Moby Dick* (1851) went so far beoynd perfection that its architectural pattern is often obscured by its architectural detail. The American novel had arrived, but unfortunately, the American reading public was far behind. Hawthorne was guaranteed a permanent niche in the tradition of the American novel when Holmes imitated him badly (*Elsie Venner*) and Henry James imitated him well. *Moby Dick* was appreciated by neither the public nor the critics, and Melville's reputation gradually declined to near oblivion. In this situation he hardly could influence the history of the American novel, not, at least, until the third decade of the 20th century, when he was revived for a period of almost hysterical adulation, and *Moby Dick* was frequently compared to Milton's *Paradise Lost*. But even if Melville had won the audience he deserved in 1851, what novelist could possibly have derived a beneficent influence from *Moby Dick*? This monumental book

remains an unexplained pinnacle in the history of the American novel if not in the history of world literature. Such works are seldom accountable in terms of literary movements.

Throughout the 19th century the sentimental novel continued to be written by "three-name women" and enjoyed by two-name women. One of these books achieved a historical importance far beyond its literary quality: Harriet Beecher Stowe unwittingly became a pioneer in the "propaganda novel" when she wrote *Uncle Tom's Cabin* (1851-2), a novel which touched the heart of the public by presenting "Pictures of slavery." A more serious propagandist arose after the war in the person of Albion W. Tourgée, whose books were social studies of the problems of Southern reconstruction, presented in a mixture of romance and realism. J. W. De Forest was also an early realist in his presentation of detail and in his satirical delineation of character, although his plots were highly romantic. A specialized branch of realism developed in the local-color school of writers, represented by George Washington Cable, Edward Eggleston, Charles Egbert Craddock, Mary Wilkins Freeman, Constance Fenimore Woolson, Thomas Bailey Aldrich, Alice Brown, Thomas Nelson Page, Sarah Orne Jewett, and Kate Chopin. Usually they portrayed the decay of older ways of life after the Civil War, either the obvious disintegration of Southern society, or changes in the economic patterns of the North.

The local-color school of the Far West, represented by Bret Harte, had long mingled with the tradition of cracker-box humor, characterized by the "tall tale." Out of these traditions sprang America's greatest popular novelist, Mark Twain, who eventually transcended all schools, colors, and regions. He contributed little to the technical development of the novel, for in studied technique he was woefully deficient. His genius could not be imitated, for it was compounded of the remarkable personality of the man himself. Imitations of his devices for securing humor fall flat when they are tried by smaller men. It is difficult, therefore, to appraise his effect on the novel. He may have had more influence than is generally recognized simply as a liberating force. By daring to ignore the literary language and mores of tradition, by challenging the platitudes of both popular and intellectual ideology, by turning sacred cows into comic cows, he may have given American literature a new vitality, and he may have inspired in the young writers of the following generations a courage to experiment in ways that were alien to Twain himself. Of his many novels, only *Huckleberry Finn* and perhaps *The Mysterious Stranger* reach some kind of perfection of the novel form. T. S. Eliot has pointed out that *Huckleberry Finn* is held together by the symbol of the river, which was doubtless a literary accident. But if he did nothing else for American literature, Twain raised humor to the level of great art, and all later humor may stem, to some degree, from him insofar as it is seriously treated as an art.

In marked contrast to Twain stands Henry James, whose influence, far from liberalizing, was beneficially restrictive. More than any other author, James improved the structural quality of the American novel, by exploring every architectural possibility, and by serving as a distinguished example. No American novelist has had a greater influence on so large a number of other great writers, American or European, for his disciples include not only Edith Wharton, Elinor Wylie, Ellen Glasgow, Willa Cather, F. Scott Fitzgerald, and Jean Stafford in America, but also Ford, Conrad, Proust, Gide, and Virginia Woolf in Europe. It well may be that the contiguous appearance of the two different masters, Twain exploiting the complete personality, and James studying the most subtle functions of craftsmanship, provided an invaluable background for the large number of 20th-century apprentices who would treat the novel more seriously than it had ever been treated before in the United States.

Twentieth Century: The new tone at the turn of the century was "Realism" (q.v.). William Dean Howells had been advocating and practising a pleasant Realism for many years, and his disciple, Hamlin Garland, went so far as to reveal, in his fiction, that a barn might not be devoid of unsavory odors. The new Realists, however, had been finding their examples in Flaubert, de Maupassant, and Balzac, rather than in Howells. In answer to Howells' statement that our novelists "concern themselves with the more smiling aspects of life, which are the more American," Frank Norris presented the brutal *McTeague* (1899), and Stephen Crane pictured what could happen in America to *Maggie: A Girl of the Streets* (1893). Howells himself was sensitive to social injustice, and he was an early champion of Crane. From a critical point of view, however, he felt that Maggie and McTeague were the exceptions and not the representatives of American life.

Although Howells was probably right, exceptional characters were not to be the exception during the next century. The closest followers of the Howells type of Realism are to be found among those novelists who delineated farm and rural life, with a particular regional setting. Willa Cather, Ruth Suckow, O. E. Rölvaag, and Mari Sandoz, among many others, wrote stark novels about immigrant farmers struggling to make a life and a living for themselves in pioneer farm situations. Even these novels, however, seemed to have a flavor of the social consciousness which is generally termed "Naturalism" (q.v.), for the land that was left for the immigrants was of poor quality and they were presented as essentially underprivileged characters. Willa Cather extended her talents to cover a much broader cross section of Midwestern society, and her success in the Realism of the average may have been due to the fact that, like Howells, she had a remarkable ability for making the average reality interesting. It is an open question how many of the Naturalists were motivated by a compulsion to correct social evil, and how many recognized the dramatic potential of social evil.

Whatever their motivation, an increasing number of novelists focussed on the underprivileged member of society, and exposed the role that society itself played in driving good people to evil, and its hypocrisy throughout the process. Reality implies honesty, and if the Naturalist novelists were to present reality, they could cut through the social platitudes and examine American life with the honesty of a scientist. Zola's original "naturalist manifesto" was predicated upon the scientific examination of life, and if later Naturalists were somewhat deficient in scientific objectivity, they retained scientific honesty insofar as they exposed social dishonesty. Dreiser's *Sister Carrie* (1900) casts society in the role of villain, and so does *An American Tragedy*, which he wrote 25 years later. Upton Sinclair was a militant leader of the muckraking movement which exposed the sad ethics of big business. Sherwood Anderson presented the small Midwestern town as a community of "grotesques." Dos Passos invented techniques to present a horizontal cross-section of New York City. James T. Farrell worked with sociologists at the University of Chicago before conceiving his social portraits of Chicago's South Side. Angered by the inhuman treatment of transients from the dust bowl of Oklahoma, Steinbeck stamped out the vintage of his own *Grapes of Wrath* (1939), which marks the zenith of Naturalist expression.

Other novelists, who are not classed as Naturalists, were also examining society with a critical eye. Sinclair Lewis documented the mores of the small town and of the Midwestern middle class with a satire so merciless and persistent that it even became monotonous. Hemingway found a dramatic potential for Realism in foreign and adventurous situations, where good people were victimized by civilization. Fitzgerald appeared to break with the Naturalists completely when he undertook his delineation of the rich, and it may be noted that this heresy caused him to be underrated for two decades, but he still examined society with judgment. Thomas Wolfe endeavored to apprehend the impression of his own life, and portrayed a society insensitive to genius. The women novelists, Edith Wharton and Ellen Glasgow, and later Jean Stafford and Carson McCullers, usually were more objective than the men in their examinations of society, but a criticism was frequently implied. The major break from the 19th century was in judging Nature or society or even God rather than the individual human being. Direct editorializing was abandoned, but the reader finds no difficulty in discovering the editorial opinion of such militant Naturalists as Farrell and Steinbeck, or the even more extreme Albert Maltz.

The danger of interpreting literary history in terms of movements lies in a tendency toward over-simplification. Insofar as the 20th-century American novel is anti-romantic in its conception, style, and appeal, it may be classified as realistic. Representation of an imperfect society implies social criticism, whatever the intentions of the author may be. Beyond the discernible movements, however, the novel was actually breaking out in every direction. Faulkner was called a Naturalist by critics who perceived only his monumental documentation of the decaying South, but his development of this material was predicated upon artistic requirements, and almost never upon scientific or propagandistic intention. Santayana's *The Last Puritan* is a timeless narrative that can not be catalogued. Saroyan expresses more whimsy than social indignation, and even Steinbeck could portray the *paisanos* of *Tortilla Flat* with love and humor. H. L. Davis perpetuates the Twain tradition of frontier humor with symbolic overtones. Several novelists have mixed either humor or fantasy with their realism: James Branch Cabel, Robert Nathan, and J. D. Salinger, among many others. Even the novels which interpreted the last war often mixed humor with tragedy, as in Thomas Heggen's *Mister Roberts*.

Meanwhile, a bountiful supply of popular romantic novels, most of them historical, has continued to appear on the market, represented at their best by Hervey Allen's *Anthony Adverse*, Margaret Mitchell's *Gone With the Wind*, and Kenneth Robert's *Northwest Passage*. Other popular novels have followed the realistic style of Hemingway, avoiding the slightest hint of decoration or sentimentality. This is especially true of the mystery novels, many of them being constructed with superb craftsmanship and true psychological characterization.

Apparently neither the critics nor the writers have decided what course the novel will take in the second half of this century. Ralph Ellison has stated that he went back to several 19th-century models in writing *The Invisible Man* (1952), and Jean Stafford seems to derive considerable influence from Henry James. Hemingway achieved his greatest structural success with *The Old Man and the Sea* (1952). Steinbeck appeared to be reaching for a greater universality in *East of Eden*. Naturalism is definitely waning, and there is a tendency to subjugate Realism to the demands of the art of the novel. Meanwhile, the earlier Realists may still have their influence on the style and attitudes of the younger writers, as when Carson McCullers borrows technical devices from Sherwood Anderson. None of the contemporary novelists shows any tendency to return to the 19th-century platitudes about this best of all possible worlds. It must be remembered, of course, that the few great novelists of that period seldom accepted the platitudes of their general public.

The advent of paper-back publishing in cheap editions, the competition from new technical mediums of narrative, the almost morbid hunger of the contemporary reader for information rather than apprehension—all of these factors will indubitably have their effect on the course of the novel. At this point in time it can only be said that the serious novelists appear to weigh their problems with great deliberation, and to avoid easy paths into novelty. The novel may be on the verge of replacement by drama, or it may be on the threshold of a great renascence.

NYE, Bill (Edgar Wilson Nye) (1850-96), born in Maine, reared in Wisconsin, went to Wyoming in 1876, where he was admitted to the bar (the examination being a mere formality). He edited the Laramie *Daily Sentinel* and wrote articles for the Cheyenne *Sun* and the Denver *Tribune*. In 1881 he founded the Laramie *Boomerang*, and the humorous sketches he wrote for it were widely reprinted and won him an international fame. They were collected in *Bill Nye and Boomerang* (1881), followed by *Forty Liars and Other Lies* (1882), *Baled Hay* (1884) and others. He used the stock humorous devices of the time, misquotation, punning, and deformed syntax, while avoiding the device of misspelling. John Dewey praised him highly for exposing pretense and superstition. He moved to New York in 1886, for reasons of health, and continued to produce satire for the *World*, while lecturing with James Whitcomb Riley.

O

O'BRIEN, Edward J[oseph Harrington] (1890-1941) edited an annual *Best Short Stories* of the years, 1914-40, and the *Best British Short Stories* during his residence in England, from 1922 until his death. He was succeeded in this enterprise by Martha Foley, with whom he had worked closely in defining the short story. (See SHORT STORY).

O'BRIEN, Fitz-James (c. 1828-62), came to America from Ireland in 1852, after wasting an £8000 inheritance in London and Paris. Entering the Bohemian life that centered around Pfaff's, he wrote editorials, verse, and stories that provided him periods of luxurious living interspersed with periods of poverty. Appointed a lieutenant and cited for gallantry during the Civil War, he died of wounds that caused a tetanus infection. Some of his stories, written in the tradition of Poe, won him a minor reputation, particularly "The Diamond Lens" (*Atlantic*, Jan., 1858), a story about an inventor whose powerful microscope enabled him to see a tiny female in a drop of water, with whom he fell in love.

O'CONNOR, William Douglas (1832-89), journalist and civil servant, is remem-bered for his defense of Whitman, *The Good Gray Poet* (1866), written after Whitman was dismissed from the Department of the Interior. He was Whitman's intimate friend in Washington and his perpetual champion. His other work includes an abolitionist novel, *Harrington* (1860); *Three Tales* (1892), containing the tribute to Whitman first published in *Putnam's Magazine* (Jan., 1868), "The Carpenter"; and Baconian pamphlets questioning the identity of Shakespeare.

ODELL, Jonathan (1737-1818), a New Jersey surgeon and Anglican minister who wrote poems satirizing and attacking the American cause during the Revolutionary War, collected in *The Loyal Verses of Joseph Stansbury and Doctor Jonathan Odell* (1860).

ODETS, Clifford (1906-), grew up in the Bronx, became an actor at 15, performed on radio programs, worked as an announcer and gag writer, and toured with a stock company. In 1928 he joined the Theatre Guild, and in 1931 he was a founder of the Group Theatre, which produced his *Waiting for Lefty* (1935), an experimental play which pretended that the audience was composed of members of a taxicab drivers' union attending a meeting. They were all waiting for Lefty to help them settle the question of whether to strike. When the play became an immediate success, his longer work about a poverty-stricken Jewish family, *Awake and Sing,* was produced the same year. His other plays include *Paradise Lost* (1935); and *Golden Boy* (1937), concerning a young violinist who wins fame as a pugilist. These dramas established Odets as a leading proletarian playwright, working in the collective ideology. Later he became a Hollywood scenarist.

O'HARA, John [Henry] (1905-   ), the eldest of seven children in a Pennsylvania physician's family, attended Fordham Preparatory School, and held a number of laboring jobs before he became a feature writer in New York and a scenarist for Paramount Pictures. His first novel, *Appointment in Samarra* (1934), is an unobtrusive example of the Zola technique of determinism, applied to a cross section of people in a small Pennsylvania town. O'Hara writes easily, seldom revising, and his skill is such that he also reads easily. *Butterfield 8* (1935) is a story of the underworld, based on an actual murder case. *Pal Joey* (1940) is about an aging night club singer, and the novel presents his letters written to a band leader. The book was successfully dramatized as a musical comedy in 1940, and enjoyed a recent revival. O'Hara's short stories, many of them appearing first in *The New Yorker,* usually are character studies, and have been collected as *The Doctor's Son* (1935), *Files on Parade* (1939),

Pipe Night (1945), and *Hellbox* (1947). *Sweet and Sour* (1954) is a collection of sophisticated comments on books and people.

O'HARA, Theodore (1820-67), Kentucky editor, colonel in the Confederate Army, and cotton merchant, wrote "The Old Pioneer," and "The Bivouac of the Dead," popular anthology poems which are frequently quoted on military tombstones and cemetery gates.

O. HENRY. See Henry, O.

OLDSTYLE, Jonathan, pseudonym of Washington Irving (q. v.).

OLDSTYLE, Oliver, pseudonym of J. K. Paulding (q. v.).

O'NEILL, Eugene [Gladstone] (1888-1953), spent his first seven years on the road, as his father starred in the company playing *The Count of Monte Cristo* throughout the United States. Eugene attended Catholic schools and Betts Academy in Connecticut, and in 1906 spent one year at Princeton. He worked in a New York mail-order house, prospected for gold in the Honduras, where he contracted malaria, and returned home to serve as assistant manager of his father's company. Next he worked his way to Buenos Aires, worked there in offices and in a packing plant, shipped to South Africa and back as mule-tender on a cattleboat, was a beachcomber in Argentina, and finally returned to New York to become an able-bodied seaman in the transatlantic service to Southampton. He next tried acting in his father's company and reporting for the

New London (Conn.) *Telegraph,* but in 1912 the discovery that he had incipient tuberculosis forced him to retire to a sanitarium. The enforced rest, for his mind as well as his body, apparently gave him the opportunity to determine what he wanted to do with his life. He began writing one-act plays, and in 1914 he enrolled in the drama workshop of George Pierce Baker (q. v.) at Harvard. From 1916 to 1919 the Provincetown Players produced several of his one-act plays, including *Bound East for Cardiff* and *The Moon of the Caribbees,* and *The Smart Set* published three of them. When he was awarded the Pulitzer Prize for *Beyond the Horizon* (1920), he received the additional recognition he deserved as a creative dramatist. *Anna Christie* won the Pulitzer Prize again for the 1921 season. New plays were produced in rapid succession: *The Emperor Jones,* 1920; *The Hairy Ape,* 1922; *All God's Chillun Got Wings* and *Desire Under the Elms,* 1924; *The Fountain,* 1925; *The Great God Brown,* 1926; *Lazarus Laughed,* 1927; *Marco Millions* and *Strange Interlude* (Pulitzer Prize), 1928; and the trilogy, *Mourning Becomes Electra* in 1931. Although O'Neill was a successful Broadway playwright, he constantly experimented with new theatric and thematic devices that one would suppose to have alienated the popular audience. In *The Hairy Ape* he employed an expressionistic technique for a naturalistic theme. In *Strange Interlude* he used the stream-of-consciousness device of the novel in a production so lengthy that the audience went out to dine between acts. In an O'Neill play

the characters might wear symbolic masks or speak their inner thoughts as an "aside." Incest and adultery frequently appeared as his surface subject matter. Almost every play was infused with a poetic symbolism. But neither poetry nor experimentation violated his shrewd dramatic intuition and skill, and if the audiences did not perceive his total intentions, they at least found an evening with an O'Neill play to be a valuable experience. In 1936 he was awarded the Nobel Prize for literature, which signified the international recognition of his stature. His final play, *The Iceman Cometh* (1946), symbolically portrayed disillusion and the approach of death, which had been haunting O'Neill through years of illness. The Iceman came for him in 1953.

OPTIC, Oliver, pseudonym of W. T. Adams (q.v.).

O'REILLY, John Boyle (1844-90), an Irish patriot, was sentenced to imprisonment in Australia for his Fenian activities, but escaped to America on a whalin ship in 1869. He joined the staff of the Catholic paper, *Pilot,* in Boston, and later became its editor and part owner. His *Moondyne* (1870) is a powerful story of convict life in Australia, and his ballads were published as *Songs from Southern Seas* (1873) and *Songs, Legends, and Ballads* (1878). He was a popular lecturer, and received honorary degrees from Notre Dame and Georgetown Universities.

ORVIS, Marianne Dwight (1816-1901), wrote *Letters From Brook Farm, 1844-*

1847, which were edited and published in 1928 by Amy Louise Reed. Her objective and optimistic reports concerning the communal project are a valuable source of information.

OSBORN, Laughton (c. 1809-78), a New York dabbler in the arts who published several books of poetry at his own expense and wrote plays that were not produced. The unfavorable reaction of the critics to his novel, *Sixty Years of the Life of Jeremy Levis* (1831), caused him to engage in a lifetime crusade against critics and authors in general.

O S G O O D , Frances Sargent [Locke] (1811-50), is remembered for her friendship with Edgar Allan Poe, rather than for the sentimental verses he praised with uncritical extravagance: *The Casket of Fate* (1840), and *The Poetry of Flowers and the Flowers of Poetry* (1841).

O'SHEEL (Shields), Shaemas (1886-), New York poet who wrote the collection, *Jealous of Dead Leaves* (1928), in the tradition of the earlier Yeats and the Irish renaissance.

OSTENSO, Martha (1900-), was born in Norway, reared in Minnesota and South Dakota, educated at the University of Manitoba, and studied the novel at Columbia University. Her first novel, *Wild Geese* (1925), describes an Icelandic farm community in Northern Manitoba with realistic perception of their mores. Her novels generally depict immigrants in the Mid-West or in Kansas, the younger generation seeking to escape the harsh domination of the parents. They include: *The Dark Dawn* (1926), *The Young May Moon* (1929), *The Waters under the Earth* (1930), *The Stone Field* (1937), *O River Remember* (1943), and *Sunset Tree* (1949).

P

PAGE, Thomas Nelson (1853-1922), son of a Virginia planter, took his LL.B. degree at the University of Virginia, and settled in Richmond to practise law. A story published in the *Century* (1884), "Marse Chan," won him a literary reputation, and in 1893 he abandoned the law for a writing career, moving to Washington. He served with distinction as Ambassador to Italy during the First World War. His stories of the pre-War South, written in the local-color tradition and employing dialect, were collected as *In Ole Virginia* (1887), *Elsket and Other Stories* (1891), *The Burial of the Guns* (1894), and *Bred in the Bone* (1904). His novels include: *Red Rock* (1898), depicting the oppressive military rule of the South after the Civil War; and *Gordon Keith* (1903), a contrast between a northern and a southern gentleman to the advantage of the well-born southerner. He also published essays, dialect verse, and a biography of *Robert E. Lee, Man and Soldier* (1911).

PAGE, Walter Hines (1855-1918), statesman, editor, publisher, novelist, letter writer, and Ambassador to the Court of St. James, was born in Cary, North Carolina. He criticized the postbellum South for its reactionary politics and urged reforms in agriculture, industry, and education. His semi-autobiographical novel *The Southerner* (1909) details his sad experiences in pressing for these reforms. From time to time he was editor of the Raleigh *State Chronicle, Forum, Atlantic Monthly,* and *World's Work.* A pioneer thinker unappreciated in his day, he is now recognized as a powerful gadfly. At the beginning of World War I he encouraged American participation even while the United States was striving for neutrality. *The Life and Letters of Walter H. Page* (3 vols., 1922, 1925) were edited by Burton J. Hendrick.

PAIN, Philip, about whom practically nothing is known including his nationality, wrote poems that resemble the work of John Donne in their religious introspection, published in Massachusetts as *Daily Meditations* (1666). He may have been the first American poet published in the colonies.

PAINE, Albert Bigelow (1861-1937), wrote the authorized three-volume biography of *Mark Twain* (1912) and edited the Twain letters (1917). He also wrote novels and plays, including *The Great*

White Way (1901), the source of Broadway's nickname.

PAINE, Robert Treat, Jr. (1773-1811), was originally named "Thomas", but to avoid confusion with "the atheist," he adopted the name of his distinguished father, a Boston socialite and signer of the Declaration of Independence. Paine edited the *Federal Orrery,* an anti-Jacobin political paper (1794-6). Bohemian living and marriage to an actress estranged him from his father and conservative friends, and although he returned to their favor for a time as a practising lawyer (1798-1803), he eventually died in poverty. His facile verse was highly regarded in his own time, his carelessness being accepted as the eccentricity of a genius. It was collected for publication in 1812.

PAINE, Thomas (1737-1809), was apprenticed to his father's trade, corsetmaking, at 13, but held various other jobs in his native England before a meeting with Franklin opened the opportunity to come to America, where he became a successful journalist. In *Common Sense* (1776) he argued with his captivating and superb logic for an immediate declaration of independence, and not only did this pamphlet assure his personal fame, but it rallied public opinion to the revolutionary cause, gave heart to Washington's army, and served as a model in substance and word for the Declaration of Independence. While serving in the Continental Army he wrote a series of 16 pamphlets in further support of the war, *The American Crisis* (1776-1783). *Public Good* (1780) opposed Virginia's claims to certain western land. He served as secretary to the Congressional committee on foreign affairs (1777-9), and traveled to France in 1781 to secure supplies. Upon his return he settled on a New Rochelle farm, where he worked on his invention of an iron bridge, and wrote against paper money inflation (*Dissertations on Government,* 1786). In 1787 he went to England to have his bridge constructed. There he was arrested for treason after issuing *The Rights of Man* (1791) in defense of the French Revolution, but William Blake smuggled him to France before the trial. He was declared a French citizen and elected to the Convention, but his middle-of-the-road policies were not welcomed by the extremists who came to power, and he was thrown into Luxembourg Prison. There he began work on *The Age of Reason* (1794-5), a deistical renunciation, not of religion, but of religious creeds. Eventually freed by Monroe's intercession, he returned to the U. S., where he was perpetually attacked for his liberal views, and died in poverty. Completely unselfish and idealistic, he bore the too familiar cross of living before his time.

PARKE, John (1754-89), a lieutenant-colonel in the Continental Army, anonymously published paraphrases of *The Lyric Works of Horace . . . to Which Are Added A Number of Original Poems* (1786); in which Roman events are replaced by American, and Augustus by Washington.

PARKER, Dorothy [Rothschild] (1893-), began her writing career with *Vogue,* and in 1917 became dramatic critic of *Vanity Fair.* Her satiric light verse was published in *Enough Rope* (1926), *Sunset Gun* (1928), and *Death and Taxes* (1931), and collected in *Not So Deep as a Well* (1936). Her stories were collected in *Here Lies* (1939). She has also collaborated with her second husband, Alan Campbell, in writing motion-picture scenarios. In addition to her reputation for her published wit, she has won a legendary reputation for witty sayings that move in the oral tradition. Her work has a teasing cynicism often derived from reversed platitudes.

PARRINGTON, Vernon Louis (1871-1929), a scholar of American literature whose *Main Currents of American Thought* (3 vols., 1927-30), an economic interpretation of American literature, affected the teaching and criticism of that discipline for a time after its publication. The first two volumes were awarded the Pulitzer Prize (1928).

PARRISH, Anne (1888-), was born in Colorado Springs, Colorado, and traveled extensively in America and abroad with her artist parents. Her novel, *The Perennial Bachelor* (1925) won the Harper prize. Her other novels include *Tomorrow Morning* (1926), *Loads of Love* (1932), *Golden Wedding* (1936), *Poor Child* (1945) and *A Clouded Star* (1948).

Partisan Review (1934-), originally partisan to the Communist party, since 1938 has championed intellectual freedom in its political articles. Literary contributors have included Dos Passos, Farrell, Waldo Frank, MacLeish, Eliot, W. C. Williams, and Wallace Stevens.

PASQUIN, Anthony, pseudonym of John Williams (q.v.).

PATCHEN, Kenneth (1911-), grew up in Ohio and went to work in the steel mills at 17. He later drifted about the country doing odd jobs. His advance-guard poems have been published in *Before the Brave* (1936), *First Will and Testament* (1939), and *Selected Poems* (1947). His first novel, showing a strong influence from Joyce, was *The Journal of Albion Moonlight* (1941), followed by *The Memoirs of a Shy Pornographer* (1945) and *See You in the Morning* (1948).

PATTEE, Fred Lewis (1863-1950), scholar of American literature, wrote *History of American Literature Since 1870* (1915), *The New American Literature, 1890-1930* (1930), and numerous other studies.

PATTEN, William Gilbert (1866-1945), using the pseudonym of Burt L. Standish, wrote more than 200 novels in the Frank Merriwell series about a fine American college boy, and many other series and dime novels for boys.

PATTON, Frances Gray (1906-), short story writer, was born in Raleigh, North Carolina. Since 1946 she has been consistently published in *The New Yorker* and other magazines. Her stories of the not always placid lives of the educated, urbane middle class in the South are written in a witty, sophisticated style. They were collected in *The Finer Things of Life* (1951). She lives at Duke University, where her husband is professor of English.

PAUL, Elliot [Harold] (1891-   ), a Massachusetts novelist, worked for American newspapers abroad after serving in the First World War. His books include *Concert Pitch* (1938), concerning expatriate Americans in Paris, and *The Last Time I Saw Paris* (1942).

PAULDING, James Kirke (1778-1860), an intimate friend of Washington Irving, with whom he collaborated on *Salmagundi* (1807-8). He also wrote the comic account of the development of the American colonies, *The Diverting History of John Bull and Brother Jonathan* (1812). He satirized Scott in *The Lay of the Scottish Fiddle* (1813). He continually wrote satires on England and English writers, as well as tales of the frontier, Virginia, and New York that showed his Jeffersonian principles. *The Lion of the West* (1830) is one of the first comedies about a woodsman in New York. His serious work includes *The United States and England* (1815) and *A Life of Washington* (1835). He was Secretary of the Navy under Van Buren.

PAYNE, John Howard (1791-1852), a New York actor, playwright, producer, and editor, wrote a long series of melodramas which were produced both in America and in England, including *The Maid of Milan* (1823), operatized by Sir Henry Bishop, and remembered for the heroine's song, "Home Sweet Home."

He and his friend Irving collaborated both in the writing of plays and in the courting of Mary Wollstonecraft Shelley, widow of the poet.

PEABODY, Elizabeth Palmer (1804-94), considered to be the prototype of Miss Birdseye in Henry James' *The Bostonians*, worked with Margaret Fuller for social reform, and kept a bookshop which was a favorite meeting-place of the Transcendental Club.

PEABODY, Josephine Preston (1874-1922), was influenced by Moody while she was a student at Radcliffe (1894-6) to try her hand at poetic drama. The plays she wrote in this medium were *Fortune and Men's Eyes* (1900), about Shakespeare; *Marlowe* (1901), a tragedy in five acts; *The Wings* (1905); and *The Piper* (1910), a drama identifying Christ with the Pied Piper who is loved by the children and cast out by adults. It won the Stratford-on-Avon prize competition and was produced in England as well as in America. Miss Peabody also contributed to American letters by encouraging Edwin Arlington Robinson when he desperately needed friendship. Her work was published in *Collected Plays* and *Collected Poems* in 1927. She was married to Professor L. S. Marks of Harvard.

PECK, George Wilbur (1840-1916), is best known for his *Peck's Bad Boy and His Pa* (1883), and other novels in the same pattern. His comic newspaper sketches in Irish dialect were published as *Adventures of One Terrence Mc-*

Grant (1871). His work was so popular that he was elected mayor of Milwaukee and governor of Wisconsin.

PERELMAN, S[idney] J[oseph] (1904-), scenarist and humorist, contributes witty sketches bordering on fantasy to *The New Yorker,* and has published several humorous works, including *Strictly from Hunger* (1937), *Look Who's Talking* (1940), *Crazy Like a Fox* (1944), *Westward Ha* (1948), and *Listen to the Mocking Bird* (1949).

PERRY, Bliss (1860-1954), English professor at Princeton and Harvard, edited the *Atlantic Monthly* from 1899 to 1910. A scholar of American literature, he published biographies of Whitman and Whittier, as well as novels and essays. His autobiography takes its title from Chaucer's description of the Clerk, *And Gladly Teach* (1935).

PETERKIN, Julia [Mood] (1880-), mistress of a South Carolina cotton plantation, became intimately acquainted with the lives of the Negro workers, and began writing about them at 40, when her son left for school. *Scarlet Sister Mary* (1928) was awarded the Pulitzer Prize, and was dramatized with Ethel Barrymore in the leading role (1929). *Black April* (1927) is another novel, concerning a giant Negro and his children, legitimate and otherwise. "The Diamond Ring" was an O. Henry Memorial Prize Story in 1930.

PHELPS, William Lyon (1865-1943), professor of English at Yale (1892-1933), in addition to academic publications on the English romantics, wrote a column for *Scribner's,* "As I Like It," which pioneered the present interest in the humanities as a popular study.

PHILLIPS, David Graham (1867-1911), New York journalist who wrote muckraking articles, essays, and 23 novels concerning social problems. His most important contribution to the novel form is *Susan Lenox: Her Fall and Rise* (1917), the story of an essentially noble minded girl who suffers from the prejudice against her illegitimate birth, and after a series of hardships, sells herself to provide medical care for a friend. She eventually discovers that prostitution on the higher social levels can provide the only escape from complete poverty and social rejection. His other novels, attacking civic corruption as well as distorted social mores, include: *The Great God Success* (1901), *The Deluge* (1905), *The Conflict* (1911), and *The Price She Paid* (1912). Phillips was murdered by a maniac.

PHOENIX, John. See George Horatio Derby.

PIATT, John James (1835-1917), Ohio journalist, and United States Consul in Ireland (1882-1893), wrote conventional verse that sometimes showed originality, at least in the themes. With Howells he wrote *Poems of Two Friends* (1860) when they were young newspaper men in Columbus.

PIATT, Sarah Morgan [Bryan] (1836-1919), the wife of John James Piatt (q.v.) grew up in Kentucky. Her 15 volumes of poetry are forgotten today, although they were less conventional than the work of her contemporaries and show some taste. She was compared in her day with Elizabeth Barrett Browning.

PICKERING, John (1777-1846), son of the prominent Revolutionary statesman, Timothy Pickering, although a lawyer by profession, learned 20 languages during his travels, and is now remembered for his avocation, philology. His *Comprehensive Lexicon of the Greek Language* (1826) became a standard dictionary, and he was the first scholar to make a study of Americanisms: *Vocabulary or Collection of Words Which Have Been Supposed to be Peculiar to the United States of America* (1816).

PIERPONT, John (1785-1866), first a lawyer and later a Unitarian minister in Boston, wrote *The Portrait* (1812), a Federalist poem; *Airs of Palestine* (1816), in praise of sacred music; and *The Anti-Slavery Poems of John Pierpont* (1843). His propagandistic sermons alienated his congregations, and he later became a Treasury Department clerk. The financier, J. P. Morgan, was his grandson.

PIKE, Albert (1809-91), son of a Boston shoemaker, became a teacher at 15, but in 1831 left for the West in search of opportunity. He became a prominent Arkansan lawyer and journalist, and wrote unusually vigorous *Prose Sketches and Poems, Written in the Western Country* (1834). He served in the Mexican War and commanded Indian troops in the Confederate Army. In his later years he devoted himself to writing poetry as well as prose on Freemasonic ritual.

POE, Edgar Allan (1809-49), was born to a Georgia lawyer turned actor and an English actress, who both died in 1811, and the boy was reared by a wealthy Richmond merchant, John Allan, from whom he took his middle name. The Allans took him to England (1815-20) where he attended school at Stoke Newington, which he described in his story, "William Wilson." A handsome and brilliant lad, skilled in boxing and swimming, Edgar was the pet of Mrs. Allan, but not of his foster father. When Edgar entered the University of Virginia in 1826, he followed the traditional pattern of the southern gentleman in sowing a few wild oats and debts in company with his fellows. Allan used this as an excuse to shift him from college to trade. Poe rebelled, and enlisted as a private in the army. That same year, 1827, he published *Tamerlane* anonymously at his own expense. In 1830, Mrs. Allan begged her husband from her death bed to rescue their foster son, which he did with the unhappy solution of procuring for him an appointment at West Point. Poe accepted only to secure a reconciliation with the merchant. When it became apparent that this was Allan's program for ridding himself of a burden, Poe deliberately got himself expelled. Before the West Point misadventure he had published *Al Aaraaf* (1829) in Baltimore, and afterward he published *Poems by Edgar A. Poe* (1831) in New York, which contained "To Helen," "Israfel," and "The City in the Sea." But this edition, like the former two, attracted no critical or popular recognition. Poe next settled in Baltimore where he found a substitute mother in his aunt, Mrs. Maria Clemm. He won $50 in a newspaper contest for "MS Found in a Bottle," and was prevented from also winning the poetry contest by their reluctance to award two prizes to the same author. In 1835 he married Mrs. Clemm's daughter, Virginia, then not quite 14, and took an editorial position with the *Southern Literary Messenger* in Richmond. The enigma of this marriage has never been solved, but it is probable he wished merely to join the family together legally, and that his young wife remained a maiden. In 1838, after failing to find a position in New York, he moved with his family to Philadelphia where he served as co-editor of *Burton's Gentleman's Magazine* (1839-40) and later as literary editor of *Graham's Magazine* (1841-2). Wherever he served, Poe contributed stories now famous in our literature, and he consistently increased each journal's circulation. *Burton's* published "The Fall of the House of Usher," and *Graham's* had the benefit of his "The Masque of the Red Death," "The Murders in the Rue Morgue," and "A Descent into the Maelström." But he was dismissed from each position for drinking, or rather, for his inability to drink, for a single glass intoxicated him. Not that he was without his reasons for drinking: editorial pay was insufficient to maintain his household, and free-lance work drew only a few dollars, often paid six

months after acceptance. The family lived in poverty, and Virginia, already ill and suffering her first hemorrhages, was not even afforded the comfort of heat in winter. She died in 1847, after they had moved to New York, in a house at Fordham, then far out in the country. In New York Poe worked for the *Evening Mirror* and even became owner of the *Broadway Journal*—too late to save it from bankruptcy. He now courted and became engaged to several wealthy widows in the hope of owning his own magazine, but complete despair and near insanity was upon him. As a critic with the *Mirror* he attacked Longfellow, and later all of "The Literati" of Boston. In 1849 he was traveling to New York to escort Mrs. Clemm back to Richmond for his wedding with a former neighbor of the Allans, Mrs. Shelton. He stopped in Baltimore, became intoxicated, and five days later was found in a gutter, apparently the victim of an election gang who made a practice of using alcoholics to vote in every polling booth for their candidate. Taken to a hospital, he died four days later, and was buried in a Baltimore cemetery beside his wife. The Rev. Rufus Griswold, a friend during his employment in Philadelphia, was named his literary executor, but served the office more as a literary execrator, for his unsympathetic and inaccurate biography led to a false conception of the poet that persisted for half a century in America. Poe's great influence was to be on the literature of France, where he was recognized by de Goncourt for the intellectual appeal of his prose, and where he exerted an effect on the French *Symbolistes* that would return through the vehicle of T. S. Eliot as an influence on 20th-century American poets. He is now acknowledged to be the founder of the detective story, and some critics even name him the originator of the modern short story. Despite his rationalizations of his own idiosyncracy in "The Philosophy of Composition" and "The Poetic Principle," he was the most perceptive critic of his period. For a time it was fashionable in schoolroom jargon to dismiss his poems as mere exhibitions of melody, without intellectual content, but as later critics unveiled the paucity of real intellect displayed by his more popular contemporaries, readers came to see in Poe's work the passion of an intellectual mind in emotional crisis.

Poetic justice, a term used to indicate situations in literature where the characters are rewarded or punished according to their just deserts.

Poetry: A Magazine of Verse (1912-), was founded by Harriet Monroe in Chicago to provide a publication devoted exclusively to the presentation and criticism of poetry. Avoiding any expressed allegiance to the poetic fashions, Miss Monroe published a wide variety of poems, both conventional and experimental. The present editor is Karl Shapiro.

POETRY IN AMERICA:
Early New England: The first book issued in the English colonies of North America was *The Whole Booke of Psalmes Faithfully Translated into Eng-*

lish Metre (1640), more commonly known as the "Bay Psalm Book," by Richard Mather, John Eliot, and Thomas Welde. The first "best-seller" in America was Michael Wigglesworth's account of *The Day of Doom* (1662). These works, and most of the Puritan rhymes, may be classified as verse, but only by an occasional stretch of the imagination may they be defined as poetry. The Puritans distrusted the sensuous and the emotional, and they conceived of art only as a functional device for preaching Calvinism. Even Anne Bradstreet (q.v.), who showed a genuine poetic ability in some of her verse, and who dared to write in behalf of women's power of reason, was too overwhelmed by doctrine to achieve any real distinction. The greatest colonial poet was Edward Taylor (q.v.), a Puritan minister who was so conscious of his deviations from orthodoxy that he requested his manuscripts be kept from the printer, and they were so kept from his death (1729) until 1937. Taylor's career parallels that of John Donne in many respects, for Taylor wrote in the metaphysical style, praised God's glory and woman's beauty, and was a divine. The fact that his verse resembled that of the Anglican, Donne, might have been sin enough for the Puritans, and they certainly would not have condoned his poems that praise God, which resemble the work of Gerard Manley Hopkins in their emotional devotion.

The Colonial South was much more in the classic tradition of poetry, and southern writers exhibited the classical virtues of decorum and style as well as the classical vice of imitation. Dryden and Pope were the models. An anonymous poem in the Burwell Papers (pub. 1814) celebrates the rebellion of Nathaniel Bacon against Governor Berkeley of Virginia (1676) in a formal elegy which invokes Mars and Minerva, and refers to Caesar and Cato for comparisons. It is probable that a number of fairly good poems were written in Virginia and Maryland during the early colonial period, but practically no work survives. William Parks established a press at Annapolis in 1726, which he later moved to Williamsburg. There he founded the *Virginia Gazette* (1736), in which he occasionally published verse. While still in Annapolis, Parks published Richard Lewis' *The Mouse-Trap, or the Battle of the Cambrians and Mice*, translated from the Latin *Muscipula* by Edward Holdsworth, one of the most notable of early literary accomplishments. He also published J. Markland's *Typographia, an Ode on Printing* (1730), and a propaganda piece written by Governor Gooch with light humor to popularize his inspection of tobacco: *A Dialogue Between Thomas Sweet-Scented, William Oronoco, Planters and Justice Love-Country, Who Can Speak for Himself* (1732). Two other poets of the South were William Dawson, a philosophy professor, and James Sterling, an Anglican rector in Maryland. Superior as the Southern poetry was in polish and prosody to that of New England, it was uninspired, and none of the Southern poets produced anything so memorable as the work of Edward Taylor.

The Revolutionary Period: Pennsylvania displayed talent in the art of translation during its early history, David French translating Anacreon (1718-30), and James Logan, Secretary to William Penn, rendering Cato and Cicero into English for publication by Benjamin Franklin. The College of Philadelphia was the scene for the first "school" of poets, which developed at mid-century under the inspired encouragement of Provost William Smith, and included Francis Hopkinson, who later became famous as a composer, Benjamin West, who would become a distinguished painter, the dramatist Thomas Godfrey, and Nathaniel Evans. The group was a precursor of the 20th-century schools of poetry, not only in looking back to classical models, but in deploring "a climate cast / Where few the muse can relish;/ Where all the doctrine now that's told/ Is that a shining heap of gold/ Alone can man embellish" (by Evans). Their inspiration, models, vocabulary and style came from a variety of sources: classic Rome, Elizabethan England, both Herrick and Milton of the 17th century, and their contemporaries in England who were just beginning to foreshadow the romantic period with touches of Gothic emotionalism. They resembled, in fact, the average schoolboy poet for whom all studies are contemporary and pertinent, for whom the class room is more real than current events. Only Hopkinson lived and wrote into the

Revolutionary period, and he supported the rebel cause with such ballads as "The Battle of the Kegs." His transition from classicism to nationalism appeared to be complete by 1787 when he celebrated the adoption of the Constitution with an earthy reference which reminds us of Whitman: "Come muster, my lads, your mechanical tools,/ Your saws and your axes, your hammers and rules:/ Bring your mallets and planes, your level and line,/ And plenty of pins of American pine:/ *For our roof we will raise, and our song still shall be,/ Our government firm, and our citizens free.*" ("The New Roof: A Song for Federal Mechanics".)

The major poet of the Revolution was Philip Freneau (q.v.), who wrote both satirical verse in behalf of the Federalists, and romantic lyrics with a nature theme, thus serving both as an example of the spirit of the Revolution and a precursor of the coming romantic movement. John Trumbull was a master of the satiric couplet, indeed, so competent a satirist that he provoked many enmities among his victims. His more cautious *M'Fingal* (1776), a mild burlesque of the Tory position, was extremely popular, and 30 editions were published after the Revolution. Others also used the Pope style, and with better results than they achieved in their more ambitious works: David Humphreys, Timothy Dwight, and Joel Barlow, who with Trumbull were called the Hartford Wits. They were essentially conservative and aristocratic in their political philosophy, and Freneau compared them to "old defunct Tories of 1775." Not only

were their politics imitative of England, Freneau charged, but so also was their verse style. Freneau wanted a virile and indigenous national verse, but when he tried to supply the need himself he was virtually ignored. Earlier mores had demanded that poetry be functional in teaching God's word, for this was the greatest glory. A new and more practical functionalism now handicapped poetic expression, for poetry seldom flourishes where there is a great deal of physical and practical work to be done, as there was in the new and fast-developing states of America. The pattern left those who celebrated American practicality, in serious poems, with less appreciation than those who opposed it, for the intellectuals at least admired themselves, and the practical people had no time to admire the poems which did them honor. A frequent pattern in the history of world literature, this has been almost a constant pattern in American literature, with occasional exceptions appearing to prove the rule.

The Nineteenth Century was to be, in many respects, an age of poetry in America, simply because so many poets would achieve such a broad compromise between popular demands and the demands of art. The first and perhaps the best of the popular poets was William Cullen Bryant (q.v.), an extremely precocious lad who at 17 wrote his memorable "Thanatopsis" now recognized as the finest lyric to be produced in America up to that time. He was a fortuitous pioneer into the new century, romantic in his mood and love of nature, highly skilled in prosody, and just liberal enough in his religion to be acceptable to the coming generations. In many respects he was the American Wordsworth, although his influence was not so great nor his career so sustained. Other poets of the earlier period were Fitz-Greene Halleck and Nathaniel Parker Willis. Halleck was inspired, but his inspiration could not compensate for his stilted imitation of outworn models. Willis was more sentimental, and almost completely uninspired. The second major poet of the century became fairly popular, after his death, but was not in the 19th-century pattern. Edgar Allan Poe rebelled against the dogma which still accounted for the popularity of the New England poets, that verse should teach. He had little influence on his century, although a chain of influences extending to France, then to England, and back to America, has given him more importance for the 20th century. Decidedly intellectual as a critic (see **Criticism in America**) he restricted his verse to the composing of highly imaginative and colorful melodies. Perhaps the most talented versifier was Henry Wadsworth Longfellow, but his very ability to turn out innumerable poems without effort, and certainly without much thought, restricted his popularity to his own generation. After 1850, New England fairly burst out with poetic achievements, memorable but not great. Holmes and Lowell restored a certain measure of wit and intellect to poetry, and as a result, their work stands up much better in this century than that of the more popular Longfellow and

Whittier. Emerson might have achieved more distinction had he practised what he preached. The new New England poets were liberal, but in the history of American thought this was to be a transient liberalism on the road to sterner considerations and more intellectual penetrations of philosophy. Thus, the poetry was transitory, to a great degree.

The major phenomenon of the century, and perhaps in American literature, was Walt Whitman. He stood even more alone than Poe, for he attempted more, and broke not only with American tradition but with all tradition. The reason may lie partly in his lack of traditional education or academic conditioning of any kind. A printer, country school teacher, editor, carpenter, loafer, observer and lover of mankind, he wrote as if he were inventing poetry, and in many ways, he was. In the place of the artificial language of Pope, or the measured cadences of Wordsworth's "common" language, or the melodies of Poe, he substituted a free verse, loose in meter, and placed his rhymes almost anywhere except at the end of the line. He invented his own metrical devices: repetition of key words and symbols, an original arrangement of assonance and alliteration, a subtle use of caesura, a meter that was primarily trochaic and dactyllic, and a symphonic rhythm to replace the steady beat of a song. He built his best poems with an architectural structure that was too subtle for his contemporaries to perceive, and he developed symbolism to express the inexpressible. His theme was a magnificent if sometimes bombastic love of experience, a love which exhilarated him or made him sad, a love which gave importance to the most unlikely words, people, scenes, or ideas. Democracy was great because it permitted the fullest realization of life. Since everyone's life was important to him, he was the perfect poet of democracy. His major weakness was a lack of any sense of humor, which is to say that he failed in his sense of proportion and relation. When his lines do not "jell," therefore, they usually are embarrassing or funny. After he had printed the first edition of *Leaves of Grass* (1855), he continually revised and added to the first group of poems as he produced new and improved editions. Most of his changes improved the individual poems, although the later poems were not always better than the early works. He was a pioneer, and he had to forge his own way. At his best he was so good that most imitations of his work seem second-rate. The form he developed was superb for his message, or philosophy, or intuition, whatever we choose to call his general statement. It is more difficult to produce memorable comedy than tragedy, and similarly, it is almost impossible to create enduring poetry out of optimism; whereas cynicism, satire, and despair will practically guarantee the sound craftsman a name. The difference in Whitman, perhaps, was realism. When he showed optimism, it was not a platitudinous ignorance of reality; it was, rather, a deep conviction in the ultimate nobility and goodness of the common man. Longfellow could opine that "Lives of great men all remind us/ We can make our lives

sublime," but Whitman knew this for the easy lie that it was. He sensed the full tragedy, for example, of Lincoln. As a worker in Washington during the Civil War, he learned the reality of corruption. Yet, he retained his idealism. If this idealism was a ponderous, aware, search for meaning, combined with an honest Christian's passion for democratic justice, could it be expressed in heroic couplets, or a sing-song iambic pentameter in measured stanzas? If the philosophy was a wandering, probing search for elusive truth, a truth so intertwined with reality that it could not be nailed down at the end of each stanza, but needed to be held in place quickly with a hundred fingers, then he needed a wandering, probing prosody, ready to spring into action when needed, suitable for moving into slums or for "Crossing Brooklyn Ferry," ready with a hundred fingers to hold down the pieces of truth until the whole was achieved. This was what Whitman, a carpenter, was inventing over the years, while graduates of Harvard and Bowdoin lived and wrote more comfortably.

From the colonial period on, there had been demands for the truly American poet, who would sing of the new land, the new ideologies, the new dignity of the common man—even the new and virile materialism. When he appeared he was obliged to set most of the editions of his book himself, and even to peddle them himself. He was even obliged to announce, himself, that he was the prophet they were seeking, but almost no one believed him. As the early poetry had been imitative of the Augustans, practically all American poets of the 19th century, from Bryant to Edmund Clarence Stedman, wrote in the English romantic tradition. Even those who tried to apprehend the poetic implications of the romantic West, from Joaquin Miller to Vachel Lindsay, failed under the weight they carried of Victorian sentiment. Whitman obviously was an outlaw from the tradition, but even milder deviations were rejected. The most able poets besides Whitman to write after the Civil War, according to 20th-century opinion, were Herman Melville, Sidney Lanier, Emily Dickinson, and Frederick Goddard Tuckerman. Melville's ponderous brain would not permit sentimentality, but he was a victim of the period clichés, and his poems seldom "got off the ground." Tuckerman experimented with prosody, ending a sonnet with an Alexandrine line, but his material was conventional. Lanier developed a conscious theory concerning the relation of poetry to music, and wrote with a melodious musical quality, and at times, even with charm. He took a bold interest in the implications of Darwin's *Origin of Species*. Yet he was entirely romantic in his approach to ideas, never examining them with intellect, but playing with them emotionally. He had the courage to break with his time, but he was so much a representative of Victorian romanticism that he could be original only in technique, and even that originality was not constructive. We may be grateful to him, at least, for so thoroughly exploring the possibility of making music, if nothing much more, out of words.

Melville, Lanier, and Tuckerman may have become more prominent in the 20th-century study of 19th-century poetry than they deserve. They were different, they dared to challenge the didacticism of their time, and they were underrated by their contemporaries. Whitman, on the other hand, experimented with a conscious purpose and with a sound poetic philosophy. The second great poet of the 19th century, although quite different from Whitman, also experimented with courage and with a sense of how the boundaries and horizons of poetry could be extended profitably. Like Whitman, Emily Dickinson derived many of her notions about poetry from the criticism of Emerson, but no two great and new poets could be more different. Whereas Whitman sent copies of his "home made" editions to important literary men in America and abroad, Emily gave up the desire for publication when Thomas Wentworth Higginson rejected some poems with suggestions for their revision which she could not accept. While he experimented in expansion, her novelty was condensation. He wrote of everything he could perceive and was ever in search of new perceptions; she lived a cloistered life in Amherst and wrote of her personal ideas and amusements. His poetry was the apprehension of his emotional reaction to a broad life, hers an intellectual apprehension of her own narrow life. His greatest heresy was a treatment of sex too open for his time, while exalting him-heresy was impertinence, bringing God and His magnificent works down to the self to the heights of his God. Her

level of her small life, treating Him as a human father with a sense of humor and logic. Whitman generalized abstractions; she made generalizations into concrete images with an epigrammatic wit. (Death is a coachman who stops for her in his carriage.) Her intelligent study of the relation between the abstract and the concrete in poetry, which allowed her to write concretely of ideas, was her major contribution to the development of American poetry, most of her other virtues being inimitable. She could have no influence until after her death (1886), since none of her poems were published until 1890, but when volumes of the poems appeared they were immediately popular, and even the critics perceived her quality.

The Twentieth Century poets who have established new patterns have tended to resemble, in varying degrees, one of the two great 19th-century poets, either Walt Whitman or Emily Dickinson. Stephen Crane, better known for his novels, was the first poet to show the influence of Miss Dickinson. Edwin Arlington Robinson, who was two years older than Crane, probably came under her influence at about the same time. Robinson was not a mere derivation, however; he was an original and powerful poet in his own right, and he also paid the penalty of originality by waiting 25 years for recognition. He does not fit too easily into any specific pattern of poetic development, and reading Miss Dickinson's poems may have done no more than encourage him in his own originality. Like her, he always kept his fine intelli-

gence in command of the poem, and was economical in expression. Again, Puritan ancestry appears to come down to him through Emerson's Transcendentalism and Miss Dickinson's microscopic studies of both Puritan and Transcendental implications. On the other hand, his skillful use of the dramatic monologue suggests Browning. He himself was not concerned with philosophical trends or schools of poetry; he simply dedicated his life, through poverty and neglect, to the creation of poems as he himself was able and wished to write them. In form, again like Miss Dickinson, he was not a rebel. In subject, he could handle the most contemporary metropolitan experiences, or small town characters, or he could take the classic story of *Tristram* and reorganize it with psychological motivations. The psychological but sympathetic study of human beings was, in fact, his major preoccupation, and in his greatest work there were psychological overtones of "the last god going home / Unto his last desire."

The nature of influence may easily be over-simplified. The teaching of literature requires organization, and organization involves the tracing of resemblances or tendencies, which we often see as a chain of influence; yet, several influences may join in a cluster of fresh apprehensions. Miss Dickinson was widely imitated in the first two decades of this century, but her visible and acknowledged influence on 20th-century poetry is not obvious. Edna St. Vincent Millay and Sara Teasdale came under her spell, but now they appear closer to 19th-century patterns than she. Robert Frost resembles her, but he is less an imitation than a parallel product of New England individualism. Miss Dickinson's type of poetry became dominant in the second quarter of this century, but she has served more as a precedent than as a direct influence for this movement: the "new poetry" of Pound and Eliot and their disciples, who recognize her as one of their clan, had the 17th-century metaphysical poets as it conscious source. When Ezra Pound appointed himself "foreign correspondent" of Miss Harriet Monroe's *Poetry* magazine in 1912, militantly establishing an American beachhead for the Anglo-American Imagism, he was also establishing a beachhead for Emily Dickinson in the 20th century.

In 1912, the definite dichotomy between the poets who resembled Miss Dickinson and those who resembled Whitman had not developed. Both schools appeared in *Poetry* Magazine as representative of the "new poetry," simply because their work was quite different from the conventional poetry, imitative of Tennyson or Longfellow or Bryant, which still dominated most periodicals. One reason for this seeming confusion of two different types of new poetry was the prominence of free verse as an innovation. Whitman had developed free verse, yet the American champion of Imagism, Amy Lowell, was also the champion of the *vers libre* movement. Of those who now appear to be descendants of Whitman—Vachel Lindsay, Edgar Lee Masters, Stephen Vincent Benét and Carl Sandburg—only Sandburg and Masters kept their verse free. Those who now appear to be cul-

tural descendants, at least, of Miss Dickinson were also divided: Eliot, John Gould Fletcher, and Miss Lowell were fairly free; Pound and William Carlos Williams vacillated; and regular patterns, however original, were employed by Hilda Doolittle, Wallace Stevens, and Marianne Moore. The first indication of a cleavage occurred in 1914 when W. B. Yeats, visiting Chicago, said in a speech that all the great influences in art and literature derived from Paris. Alice Corbin Henderson, assistant editor of *Poetry,* wrote a reply to Yeats which defended the American tradition of the "Chicago school" represented by Lindsay, Masters, and Sandburg. Later, Amy Lowell took over Imagism, and those who did not follow her cult came to recognize Eliot as their leader, especially after he became exceedingly articulate as a critic. His disciples developed "the New Criticism," which has been a dominant influence on the poetry of the past thirty years, as well as on the criticism of poetry. The New Critics, Allen Tate, John Crowe Ransom, and Robert Penn Warren, are also poets. They have broken completely with the free verse movement, using the metaphysical poets of 17th-century England as their model. (For a more complete exposition of their theories, see Criticism in America, "The New Critics.")

The Whitman influence was not so persistent, partly because of the dominating influence of the New Critics, but also because a system which rejects disciplinary control places an unusual demand on any new poet. Without traditional disciplines to direct and sustain him, he is too dependent upon his own genius. If he lacks genius, he is apt to fail completely. Carl Sandburg appears to be the only follower of Whitman who has produced poetry of quality during the past several decades. A large number of young poets have been quite successful in following the dictates of Eliot and his disciples, however, since the new system protects them from many of the mistakes which second-rate poets are heir to.

The New Critics claim Emily Dickinson as one of their kind, and their technical policies resemble hers; however, a more likely descendant in theme as well as in style is Robert Frost, who has been independent of critical philosophies. He is also a metaphysical poet in his ability to transfigure physical into metaphysical experience, although his images are so easily grasped that his poems do not seem at all like those of the neo-metaphysical poets, an appearance that he shares with Miss Dickinson. His style, suggested by Emerson in "Monadnock," depends on his conscious effort to set the "sentence sounds that underlie the words" of common speech in the traditional meters of poetry. Thus he develops a counterpoint of sentence rhythm and verse rhythm. Like Miss Dickinson, Frost starts his poems with the common experiences of New England life, then, developing his theme more leisurely than she, he uses his physical objects as symbols, so deftly that the careless reader will miss the metaphysical intention altogether. Like Whitman, Miss Dickinson, and

Robinson, Frost worked out his theory and accomplishment of poetry in isolation, disdaining any compromise with fashion that might have won him an early audience. He has also avoided political passions, retaining the objectivity which is consistent with his true sense of irony and wit.

The most dedicated of experimenters among 20th-century poets were Ezra Pound and T. S. Eliot (qq.v.; see also Criticism in America for the history of their development). They were also the least American in their derivations, and have spent most of their adult years abroad. Eliot became a British citizen; Pound was brought back to America, during World War II, to face charges of treason, since he had neglected to legalize his Italian citizenship. Their Americanism was manifest in their reaction against Americanism. Pound was willing to dig into any culture except the American for stimuli: China, Augustan Rome, or Dante's Florence. Eliot dug with more perception and achieved fruitful results. Pound was more brilliant and less disciplined. His effect on others could be only a brilliant stimulus, except as it was filtered through the disciplined Eliot. As Longfellow and Tennyson mirrored the optimism of the 19th century, Pound and Eliot have reflected the pessimism and cynicism of the 20th century, and once they were accepted, they enjoyed a parallel sphere of influence. Their trans-Atlantic contribution to American poetry and culture has been the stimulation of a greater and more intelligent interest in traditional culture.

Eliot has been the source of a considerable influence on contemporary American poetry, in showing the strength of irony and wit and the uses of symbolism; yet none of his disciples writes like him. We know the influence has been strong, but it is difficult to realize its operation. His most deliberate disciple was Hart Crane, yet Crane sought in *The Bridge* to refute *The Waste Land* of Eliot. If Eliot provided a technical example, Whitman provided Crane's message. The technique alone could not be exploited without the strength of Eliot's philosophy. Crane was a mystic and a romantic, capable of great eloquence, but not of a sustained structure. Ransom, Tate, and Warren owe great debts to Eliot, in their poetry as well as their criticism, but they carried the intellectual consciousness to such an extreme that their poetry seems cold by comparison. The nature of Eliot's influence speaks well for his contribution, for close imitation does not make a strong poetry, and the poet who can easily be imitated lacks strength of individual quality.

The Whitman-Dickinson dichotomy serves an examination of 20th-century poetry better as an illustration than as a formula for making truthful distinctions. In the minds of many Americans it has been reduced to a distinction between those poets who are easily understood and those whose intelligence is ambiguous or obscure. For this reason, Frost was, until recently, grouped with the Whitman descendants, whereas the more difficult poets, William Carlos Williams, Marianne Moore, and Wallace

Stevens were catalogued in the "new" school, which we have identified with Miss Dickinson. Williams has more reason to be so classified than Miss Moore or Stevens; he was greatly admired by his college friend, Pound, and he was influenced early by the Imagists, but he directed his images to the homely life of common and even sordid people with a sensuous sympathy. Miss Moore was a disciple of Yeats, whom she described as "a literalist of the imagination," and she produces "imaginary gardens with real toads in them" in a mathematical verse that is counted by syllables instead of by feet. Wallace Stevens is more a descendant of Lanier than of Dickinson or Whitman, but he seeks an alliance with painting rather than music, and produces a lush, sensuous impressionism—bordering on expressionism, with the coloring of Matisse and the structure of Picasso. As with so many of our greatest poets, recognition of his rich sensibility has come late in his life; however, this was in part his own responsibility, for his best work was published late in his life.

Indeed, contemporary American poetry appears to need a reappraisal which will de-emphasize the importance of trends and schools. It may very well be decided that our greatest poets are our most independent.

The poet most isolated from trends and schools is Robinson Jeffers, who writes in Carmel, California of the oppressive luxury which the transient sickness called civilization has brought to the once clean frontier. Carrying pessimism far beyond Eliot or the new school to a Spenglerian despair of any cure for humanity, he uses such violent action as murder, incest, and rape as symbols of human ugliness. He has also carried the technical possibilities of accentual prosody beyond Whitman's quantitative experimentation. He has developed great poetic power for delivering a powerful message, but the average reader's inability to take his message seriously tends to become also an inability to take his longer poems seriously. If time dissipates our concern with message, he may emerge as one of the great poets of the century.

Poetic activity has tended, even more in this century than in the past, to receive its vitality and encouragement from the large eastern publishing centers, and a casual glance at most textbooks of literature would suggest that only Boston and New York are capable of producing poets. The southern poets, Tate, Ransom, and Warren, overcame a prejudice against regional development by organizing and publishing their own journals. Sandburg also owed his emergence, from the Middlewest, to organization in Chicago that centered around *Poetry* Magazine. Jeffers may have emerged purely through the shock of recognition. American editors rejected Frost until the English critics informed them he was great. Thousands of poets are writing with serious intention in every part of America, yet most of the current "name" poets are discovered in New York and promoted through special coteries. The very abundance of poetry being produced in America may confuse the selection of

our representative and major poets, may necessitate some special introduction to the public.

New England, the South, and the Middlewest have at least achieved poetic identities, and the Pacific Coast has produced such miscellaneous talents as Joaquin Miller, H. L. Davis, and Robinson Jeffers. The Far West, or Rocky Mountain region, like the Pacific Coast, has failed to develop a poetic identity, although every state in the region contains innumerable romantics who are busy emoting about the glory of God's grandeur as displayed in the magnificent mountain scenery. Oscar Wilde predicted that this region would be a poetic desert, because overpowering scenery overpowers poets. The West has developed a single major poet who uses western geography and history in his frame of reference and yet keeps the mountains in their place. Thomas Hornsby Ferril (q.v.) uses his knowledge of science as a discipline to protect his examination of the West from sentimentality. Actually, he is more a poet of science than of the West, for the latter is merely his focus point in objective examinations of life and time that reach backward for geological eons; but whatever his subject, it is seen from the point of view of the westerner. Like Frost, he has worked out his own theory of poetry, independent of current movements, and also like Frost, he was long assumed to be in the Whitman instead of the metaphysical camp, where he can more logically be placed. There is a growing awareness of his special and individual poetic accomplishment, and signs that he may be the next of the older poets to be "discovered." Discovery by the academic critics that the West has been disciplined might, in turn, open a path for the recognition of a few other western poets who write distinguished verse with a less emphatic regional emphasis than Ferril.

Among the youngest generation of poets, none of whom has been properly appraised, there is a tendency toward imitation of both the theme and style of their most distinguished predecessors. Such a tendency has never before produced great poetry in America. The younger poets display an amazing ability for craftsmanship, for producing successful and polished poems. There seems to be, however, no special reason for reading the poems except as an exhibition of technique, and even this exhibition loses much of its importance where skill is so plentiful. The poet's message may not be important as a message, but it needs to have some poetic significance. Today the subject of pessimism and decadence is as trite as optimism became 75 years ago, and it is repeated as often. While there is an abundance of poets, the market for poetry has considerably declined under the pressures of the practical materialism which the young poets so often deplore. Poets therefore become the audience for one another, and tend to write for one another, so that hopes of reclaiming the public attention are limited. This appears to be the major problem facing poetry in America at mid-century.

POLLARD, [Joseph] Percival (1869-1911), a German who came to the U. S. in 1885 after study in England. His criticism of sentimentality in American literature and his interpretations of the contemporary European literature had an immeasurable effect on American writers who were looking for new directions at the turn of the century. He also wrote two plays, *Nocturno* (1906), and *The Ambitious Mrs. Alcott* (1907), and the novel, *The Imitator* (1901), satirizing Richard Mansfield. Many of his critical essays were collected in *Their Day in Court* (1909).

POLLOCK, Channing (1880-1946), New York dramatist and critic, wrote innumerable farces and melodramas during the first two decades of this century, then turned to allegorical plays about *The Fool* (1922); *The Enemy* (1925); *Mr. Moneypenny* (1928); and *The House Beautiful* (1931), which failed to excite the critical acclaim he had hoped for. He retired to reminisce about *The Adventures of a Happy Man* (1939) and the *Harvest of My Years* (1943).

POOLE, Ernest (1880-1950), grew up in Chicago, and after graduation from Princeton (1902), studied and wrote magazine articles on the slum conditions in New York City. As correspondent for *Outlook* he covered the Chicago stockyards strike in 1904, while helping Upton Sinclair gather material for *The Jungle,* and in 1905 he covered the abortive Russian revolution. The *Saturday Evening Post* sent him to Europe in

1914 as war correspondent, and in 1917 he covered the Russian revolution. He wrote a long series of novels exposing poverty and relating his experiences in Russia as well as the reactions of Russian visitors to America. He also wrote character novels, including *His Family* (1917), which was awarded the first Pulitzer Prize for the novel; *Millions* (1922), a story about disappointed heirs; and *One of Us* (1934). His novel, *The Harbor* (1915) concerning a conflict betwen a union and a monopolist, was published in nine countries.

PORTER, Gene Stratton. See **Stratton-Porter.**

PORTER, Katherine Anne (1894-), a descendent of Daniel Boone, was born at Indian Creek, Texas, attended private southern schools, and became an international journalist, and a writer of novels, novellas, and short stories. Her stories were collected in *Flowering Judas* (1930) and *The Leaning Tower* (1944). *Pale Horse, Pale Rider* (1939) contains three novellas.

PORTER, William Sydney. See **Henry, O.**

PORY, John (1572-1635), English geographer, colonist, newsletter writer, and student of Hakluyt, was a Member of Parliament who in the employ of Sir Dudley Carleton went to Virginia in 1619 as secretary of state for the colony. His colorful accounts of Virginia, among the earliest news reports from English settlements, appeared in John Smith's *Generall Historie*. His description of

Plymouth Colony, edited by Champlin Burrage, was published in 1918. Other letters have been printed in recent collections.

POUND, Ezra [Weston Loomis] (1885-), was born in Idaho, but from the age of two he grew up in Pennsylvania. He studied at the University of Pennsylvania and at Hamilton College, from which he graduated in 1905. In 1906 he took his M. A. degree from Pennsylvania, and spent the following year in Europe doing research on Lope de Vega, the dissertation subject for his projected Ph. D. degree. In 1907 he taught at Wabash College, in Indiana, but was released at the end of four months, despite his extraordinary scholastic abilities, for being too unconventional. He returned to Europe, and his first book of poems *A Lume Spento* (1908) was published in Venice. He settled in London (1908-20), where a series of his books of poems were published: *Personae* and *Exultations*, both in 1909; *Provença* (1910); *Canzoni* (1911); *Riposates* (1912); and a translation of *The Sonnets and Ballate of Guido Cavalcanti* (1912). The originality and metrical experimentation in these works attracted attention, and perhaps his blonde hair and red beard helped also to identify him as an unusual figure. He became associated with the English *avant garde* movement, and soon came to dominate it. With T. E. Hulme he founded the Imagist movement (q.v.), and after Amy Lowell had taken it back to America with her, he founded Vorticism, and with Wyndham Lewis published *Blast* (1914-15), the Vorticist manifesto. He served as European correspondent for *Poetry* Magazine, and sent many experimental poems, both his own and those of his friends, to the editor, Harriett Monroe. He stimulated an interest in Chinese and Japanese verse. He championed Eliot and Joyce. He was, in fact, the prime influence on the poetry of the following decades, either in his own right or through his many and various disciples. In 1920, finding England sterile, he moved to Paris, where his influence turned more to prose, and where, with Ford Madox Ford and Gertrude Stein, he expounded his theories to Sherwood Anderson and young Hemingway. In return, Hemingway gave him boxing lessons. He moved on to Italy in 1924, where he lived in Rapallo, and briefly edited *Exile*. He was awarded $2,000 by *Dial* in 1927 for distinguished service to American literature. In 1939, during a visit to America, he received an honorary degree from Hamilton College. Meanwhile he had become interested in the relationship of economics to culture, and held that capitalism as manipulated by the Jews stimulated bad taste. During World War II he broadcast Fascist propaganda over the Rome radio, for which action he was later arrested by the American forces for treason. Considered mentally ill, he was not brought to trial, and is confined to a sanitarium. His major poetry is to be found in *Hugh Selwyn Mauberley* (1920), characterized by his conversational style and rich literary allusion, and in his brilliant and incoherent *Cantos*, of which 71 were published between 1925 and 1940. He intended to write 100 in a fugue-like structure to resemble Dante's *Comedy*, and treating the history of civilization. His best known criticism is found in *Polite Essays* (1937), containing brilliant, outspoken, tirades against conventional academic approaches to literature, mixed with prejudice, profanity, and anti-Semitism resulting from his hatred of usury.

PROKOSCH, Frederic (1909-   ), Professor at Yale and New York University, was born in Madison, Wisconsin, and educated in several American states and European countries as his father, a professor of German, took various assignments. He took his Ph. D. at Yale, writing on "The Chaucer Apocrypha." His poems have been collected in *The Assassins* (1936), *The Carnival* (1938), and *Death at Sea* (1940). Of his many novels, *The Asiatics* (1935) was one of the most successful, having been translated into five languages. *The Seven Who Fled* (1937), concerning Russian exiles, was the Harper Prize Novel in 1937.

Proletarian literature, a term applied to that writing which chronicles the suffering of the oppressed and poverty stricken members of society, usually with the purpose of arousing public indignation and effecting social improvement. Social propaganda is the Marxist designation of literature's function, a doctrine that is retained by the Communist Party, but some writers of proletarian literature are concerned rather with strengthening capitalism through the amelioration or correction of its weaknesses and misapplications, while other writers see proletarian situations as a source of dramatic material. The major authors in the proletarian movement have been Jack London, Theodore Dreiser, Sherwood Anderson, James T. Farrell, John Dos Passos, and John Steinbeck.

Pulitzer Prizes in Journalism and Letters were created through the bequest of $2,500,00 by Joseph Pulitzer, owner of the New York *World* (1883-1911), to found the School of Journalism at Columbia University, with the provision that a part of the interest from this sum be given in annual prizes to encourage public service, morals and education, as well as American literature. It is essentially a journalism prize, and has encouraged local investigations of graft as well as distinguished reporting in international affairs. The literature awards have given recognition to conventional writers with an established reputation more often than they have encouraged new or unconventional writers. Faulkner had been publishing novels for 25 years when he won the Nobel prize, but none of his novels received a Pulitzer Prize. The awards in literature are for the novel, play, poetry, biography, and history.

Puritanism was developed among the English middle class merchants in the early 17th century as a movement for the elimination of certain rituals in the Church of England. Later the movement became more radical in the changes it wished to enforce, and political in its opposition to the Stuarts, a development that culminated in the Revolution of 1640-60. During this period many Puritans settled in New England, and soon established a theocracy governing the area. (The original settlers of the Plymouth colony of 1620 were Pilgrims, and complete Separatists from the English Church.) With the Restoration of the English crown, Boston became the center of the Puritan movement, which had become more authoritarian and positive in its Protestantism than it was earlier in the century. Although it was Congregational in policy, the New England church derived its theology from the teaching of John Calvin (1509-64), and from his followers who formulated the Synod of Dort (1618-19). Predominant in Calvinism, and later in Puritanism, were two positive beliefs: (1) the *total depravity* of man after Adam's fall, and his inability to exercise free will; (2) the *unconditional election* of certain chosen individuals for salvation through God's predestination. This theological system gave the sinner no incentives for reform, since his very inclinations toward evil suggested to him that he was not one of the elect, and he could accept his total depravity as an inheritance from Adam, rather than as a personal weakness. But the Puritans were distinguished less for their logic than for the intensity of their superstitions concerning the spiritual world. Nonconformists were burned as witches, not simply as a political exigency but out of profound and

Photo: New York Times

GEORGE SANTAYANA

Photo: Philippe Halsman

JOHN STEINBECK

MARK TWAIN

WALT WHITMAN

passionate belief. Gradually the severity of the doctrine declined as the theocracy lost its supreme power. Puritan attitudes continued to dominate the mores of the New England upper class, however, and in the 18th and 19th centuries were spread to the great American middle class. These mores are characterized by intolerance, strict and illogical morality, and the identification of wealth with respectability.

Q

QUICK, [John] Herbert (1861 - 1925), Iowa novelist whose work includes *Vandemark's Folly* (1922); *The Invisible Woman* (1924); *One Man's Life* (1925), an autobiography; and a history of *Mississippi Steamboatin'* (1926, in collaboration with Edward Quick).

QUIN, Dan, pseudonym of Alfred Henry Lewis (q.v.)

QUINCE, Peter, pseudonym for Isaac Story (q.v.).

QUINCY, Edmund (1808-77), son of Josiah Quincy, prominent Bostonian, historian, and president of Harvard. Edmund Quincy became a radical Abolitionist, writing pamphlets and editing the *Non-Resistant* with William Lloyd Garrison. He wrote a biography of his father and several works of fiction, including *Wensley, A Story Without A Moral* (1854); and *The Haunted Adjutant* (1885), a collection of short stories.

QUINCY, Josiah Phillips (1829-1910), historian and poet, the fourth generation to bear the name, Josiah, received his Bachelor's, Master's and Law degrees from Harvard, but practised law only a short time before devoting himself to literature. He wrote short stories for the *Atlantic* and *Putnam's Magazine,* which were collected in *Peckster Professorship* (1888), and a dramatic poem, *Charicles* (1856).

R

RALPH, James (c. 1695-1762), born in New Jersey, accompanied Franklin to London in 1724, and never returned. The only attention he attracted with his poetry was castigation from Pope in the second edition of the *Dunciad,* but his ballad-opera, *The Fashionable Lady* (1730), had the distinction of being the first play written by an American to be produced in London. He was most successful as a political journalist, and was associated with Fielding as assistant editor of the *Champion.*

RANSOM, John Crowe (1888-   ), son of a Tennessee minister, attended Vanderbilt, and as a Rhodes Scholar spent three years at Oxford, studying the classics and mathematics. He taught at Vanderbilt from 1914 to 1937, then moved to Kenyon College, where he founded the *Kenyon Review* (1938). Earlier, in Nashville, he founded *The Fugitive* (1922-25). His poems have been published in *Poems about God* (1919), *Chills and Fever* (1924), *Two Gentlemen in Bonds* (1927), and *Selected Poems* (1945). As the leader of the *Fugitive* group (q.v.), he best exemplified the mixture of southern traditions with the metaphysical technique, achiev-

ing a blend of acrid humor, indirection, elegance and decadence. The traditionalist criticism which he writes in support of his approach to poetry has been published in *God Without Thunder* (1930), *The World's Body* (1938), and *The New Criticism* (1941). (See CRITICISM, *New Critics*).

RAWLINGS, Marjorie Kinnan (1896-), grew up in Washington, D. C., and after her graduation from the University of Wisconsin, wrote publicity, newspaper articles, and syndicated verse. In 1928 she moved to a Florida orange plantation, and devoted her time to creative writing. In 1933 she won the O. Henry Memorial award with the story, "Gal Young Un," and *The Yearling* (1938) won the Pulitzer Prize. Her other works include *Jacob's Ladder,* *South Moon Under* (1933), and *Golden Apples* (1935), all set in the Florida "cracker" country.

RAYMOND, George Lansing (1839-1929), son of a Chicago merchant, graduated from the Princeton Theological Seminary (1865) after taking his B.A. and M.A. degrees at Williams College, studied art in Europe for three years, and became a professor of English Literature and elocution at Williams. In 1880 he was appointed to the chair of oratory and aesthetics at Princeton. He wrote a novel, poems, dramas and texts, including *The Orator's Manuel* (1870), but his major work was in the study of aesthetics, published in an eight volume series *Comparative Aesthetics* (1909).

READ, Opie [Percival] (1852 - 1939), southern journalist who edited the *Arkansas Traveler* (1882-92), a humorous paper, until his satire of rural southerners made him so unpopular that he moved himself and his paper to Chicago in 1891.

READ, Thomas Buchanan (1822-72), of whom Hawthorne commented that his paintings were poems and his poems pictures, wrote many poems for the tragedian, James Edward Murdoch, to recite as the two of them traveled to raise money for the Civil War soldiers. He was regarded as one of the foremost poets in his day, and produced more than ten volumes of conventional, facile verse.

REALF, Richard (1834-78), son of a Sussex constable, was admired as a child poet by Lady Byron, who procured him a position in the household of her nephew, but his seduction of one of the daughters resulted in his immigration to America (1854). He was an aide of John Brown in Abolitionist movements, and led a Negro regiment in the Civil War. A series of marriages without intervening divorces, added to his perpetual poverty, led to his final suicide in San Fracisco.

His *Poems by Richard Realf, Poet, Soldier, Workman* (1898) are a mixture of intense passion, superb lines, bombast, and clichés.

Realism, a critical term applied to that literature which seeks to portray life and destiny as it actually happens, rather than portraying the classical tragedies of kings, or the romantic ideals of a dream life. Tolstoy, Flaubert, and Balzac pioneered this movement in 19th-century Europe, although Chaucer had used the same approach in most of his Fabliaux and in the headlinks of the Canterbury Tales. Howells was a leader of the movement in America, but many local-color writers, such as Kate Chopin, De Forest, Eggleston, Sarah Orne Jewett, and E. W. Howe, chose to represent the life they saw without deliberately adopting Realism as a movement. Many early Realists turned to *Naturalism* (q.v.), charging that Howells' Realism gave too happy a picture. Thomas Wolfe used a realistic Impressionism, and Faulkner verges on a realistic Expressionism. Most contemporary writers are essentially realistic, however they may modify a representation of actual life to gain dramatic interest. Thus Hemingway is realistic in depicting unusual and rather glamorous action; Steinbeck presents a realistic picture of people in exceptional poverty; and Fitzgerald presented the incredibly rich with accuracy. Among the poets, Frost, Sandburg, and Masters use the realistic technique, and among the dramatists, O'Neill, Rice, Williams, and Miller. The recent documentary motion pictures are outstanding examples of the attempt to strip narrative of every romantic device and to record life as investigations show it actually happened. (See also *Realism* under CRITICISM.)

Reedy's Mirror (1913-20), founded by William Marion Reedy (1862-1920) to replace the St. Louis *Sunday Mirror,* of which he had been editor. The earlier paper was devoted to local society reporting, and with the change in name, Reedy made it a literary and political review. He published Masters, Teasdale, J. G. Fletcher, Julia Peterkin, and Babette Deutsch.

REESE, Lizette Woodworth (1856-1935), Baltimore poet who, in her own individual way, pioneered the break with Victorian sentimentality in her concise, simple, personal poems, even though her subject was nature—a topic that generally brought out the worst in 19th-century Americans. Her books include: *A Branch of May* (1887), *A Handful of Lavender* (1891), *A Wayside Lute* (1909), *Wild Cherry* (1923), *Little Henrietta* (1927), and *Selected Poems* (1926).

Regionalism was a term developed to catalogue, and often to patronize, those writers who grew up outside the metropolitan centers of culture and quite properly used indigenous material for their stories, novels, and poems. After a writer becomes famous his regional interests are taken for granted, and the term is seldom applied to him.

REID, Christian, pseudonym, of Mrs. Frances Christine Fisher Tiernan (q.v.)

REMINGTON, Frederic [Sackrider] (1861-1909), attended the Yale School of Fine Arts, and during his youth worked as a western cattleman, which experience provided the background for his famous paintings and illustrations of western life. Needing a narrative to accompany his magazine illustrations, he became a writer of sketches that are collected as *Pony Tracks* (1895), *Crooked Trails* (1898), and *The Way of an Indian* (1906).

REPPLIER, Agnes (1858-1950), Philadelphia essayist, was taught by French nuns at a Pennsylvania convent. She contributed many essays to the *Catholic World* and the *Atlantic Monthly,* distinguished for their wit and scholarship. These were collected in seventeen books, including *Books and Men* (1888), *Counter Currents* (1916), and *To Think of Tea!* (1932). She also wrote bibliographical studies and autobiographical works, as well as a study of humor, *In Pursuit of Laughter* (1936).

REYNOLDS, Jeremiah N. (1799?-1858), a polar explorer and pseudo-scientific theorist, wrote an account of the whale, Mocha Dick, which Melville used in writing *Moby-Dick.* Poe used part of Reynolds's address to Congress in *The Narrative of Arthur Gordon Pym.*

RHODES, Eugene Manlove (1869-1934), grew up in Nebraska, was a cattle-man in New Mexico, and lived in New York and California in later life. As a writer of novels and short stories concerning the cattle industry he gave a realistic portrait of the western locale and

(189)

characters within his romantic narratives. His works include: *Good Men and True* (1910), *West is West* (1917), and *The Proud Sheriff* (1935).

RICE, Alice [Caldwell] Hegan (1870-1942), Kentucky writer of juvenile stories, is remembered for her *Mrs. Wiggs of the Cabbage Patch* (1901).

RICE, Elmer [Elmer Reizenstein] (1892 -), grew up in New York City, and studied law at night while working in a law office by day. His first play, *On Trial* (1914) was an immediate success, and he has been a professional dramatist ever since. *The Adding Machine* (1923) uses Expressionism (q.v.) in developing a theme concerning the replacement of men by machines. *Street Scene* (1929, Pulitzer Prize), is a tragic picture of an Irish and a Jewish family in the slums. He wrote several melodramas, sometimes in collaboration (with Dorothy Parker, Philip Barry, and Hatcher Hughes), the most successful being *The Left Bank* (1931) and *Counsellor-at-Law* (1931). He has also written novels and directed the plays of the Playwrights Producing Company, which he founded with Maxwell Anderson, S. N. Behrman, Sidney Howard, and Robert Sherwood.

RICH, Helen, born in Sauk Centre, Minnesota, became a newspaper reporter and free-lance writer in America and abroad. She finally selected Breckenridge, Colorado in the high Rockies as her home, and this is the scene of her novels: *The Spring Begins* (1947) and *The Willow-Bender* (1950). She has captured the atmosphere and the curious economic . social relations of the isolated western mining town better than any other novelist. Her realism is fortified by almost perfect structure and a delicate interplay of human emotions.

RICHARDS, Laura Elizabeth (1850-1943), author of children's books, won the Pulitzer Prize in collaboration with her sister, Maud Howe Elliott, for *The Life of Julia Ward Howe* (1916), their mother.

RICHTER, Conrad (1890-), grew up in Pennsylvania, where he became a journalist and writer of children's stories. In 1928 he moved to New Mexico, where his studies formed the basis for his novels of the southwestern frontier, including *The Sea of Grass* (1937) and *Tacey Cromwell* (1942), and a collection of short stories, *Early Americana* (1936).

RICKETSON, Daniel (1813-98), a great friend of Thoreau and other Transcendentalists, wrote *Factory Bells and Other Poems* (1873), distinguished only for sympathy expressed for the working man.

RIDGE, Lola (1871-1941), born in Ireland, lived in Australia and New Zealand, and came to the U. S. in 1907, where she became a writer for popular magazines. Her verse, showing her sympathy with the labor movement, was published in *The Ghetto and Other Poems* (1918), *Sun-Up* (1920), *Red Flag* (1927), *Firehead* (1929), and *Dance of Fire* (1935).

RIDING, Laura (1901-   ), expatriate New York poet and novelist, lived in Majorca until the Spanish Civil War drove her to England. Her nine volumes of poems were published in *Collected Poems* (1938). With Robert Graves she established the Seizin Press (1927), and also collaborated with him in making *A Survey of Modernist Poetry* (1927). *A Trojan Ending* (1937) is a novel version of the Trojan War. Her poetry is in the modern tradition of private, abstract imagery.

RIGGS, Lynn (1899-), Oklahoma cowpuncher, worked as a movie extra in New York, studied drama at the University of Oklahoma, and settled in Santa Fe. Both poet and dramatist, he has derived material from the folklore of his native state. *Green Grow the Lilacs* (1931) was transformed into the musical comedy hit, *Oklahoma!*, by Rodgers and Hammerstein, and won the Pulitzer Prize (1943). His other plays include *Roadside* (1930), *Russet Mantle* (1936), *The Cherokee Night* (1936), and *The Cream in the Well* (1941). His two collections of poetry are *The Iron Dish* (1930), and *Listen Mind.*

RILEY, James Whitcomb (1849-1916), Indiana journalist and "Hoosier Poet," wrote verse in the dialect and homespun philosophy of the people he knew in his home state. From 1877 to 1885 he contributed verse to the *Indianapolis Journal* under the pen-name of "Benj. F. Johnson of Boone," and this work was published in *The Old Swimmin'-Hole and 'Leven More Poems* (1883). Other publications were *Afterwhiles* (1887), *Rhimes of Childhood* (1890), and *Poems Here at Home* (1893). He was also a successful lecturer, and appeared with Bill Nye, reading his own poems.

RIPLEY, George (1802-80), son of a Massachusetts merchant, graduated from Harvard and its Divinity School, and became a leading, liberal Unitarian minister and editor. He resigned in 1841 to direct the Brook Farm. In 1849 he moved to New York as literary editor of Greeley's *Tribune,* and contributed also to *Harper's.* With Bayard Taylor he edited *A Handbook of Literature on the Fine Arts* (1852), and with C. A. Dana he compiled the 16 volume *Cyclopaedia* (1858-63).

ROBERTS, Elizabeth Madox (1886- 1941), Kentucky poet and novelist, graduated from the University of Chicago in 1921, and the following year published her second collection of poems (which had won the Fisk Prize at Chicago), *Under the Tree. Song in the Meadow* (1940) is a collection of her later poems. Her novels deal with rural Kentucky, and include *The Time of Man* (1926), *My Heart and My Flesh* (1927), and *The Great Meadow* (1930). Her stories were collected in *The Haunted Mirror* (1932) and in *Not by Strange Gods* (1941).

ROBERTS, Kenneth [Lewis] (1885-), grew up in Maine, and graduated from Cornell. Before the first World War he was a journalist, and afterwards he spent 9 years as a traveling correspondent for the *Saturday Evening Post*. In 1928 he retired to Italy to write the first of his many historical novels, *Arundel* (1930), concerning Benedict Arnold's march on Quebec. *Northwest Passage* (1937) finds a hero in Robert Rogers, who attempted to discover a Northwest passage. *Oliver Wiswell* (1940) chronicles the Revolution through the eyes of a Loyalist soldier. *Lydia Bailey* (1947) concerns the Tripolitan War. Roberts thoroughly masters the history for each novel with diligent research, sometimes taking two years for the preparation before beginning the composition. He is an accurate historian who prefers to present his history in the form of fiction.

ROBINS, Elizabeth (1865-1952), American actress and novelist, who became famous in London for her playing of the feminine lead in Ibsen's plays. More important than her many novels are her recounting of *Ibsen and the Actress* (1928); and *Theatre and Friendship* (1932), containing the letters Henry James wrote to her. Her novels include *The Magnetic North* (1904), concerning the Klondike, and *Come and Find Me* (1908).

ROBINSON, Edwin Arlington (1869-1935), a youngest son in a family of declining fortune, grew up in the "Tilbury Town" of his poems, Gardiner, Maine. At seven he was reading Shakespeare; at eleven he was writing poetry; and while still in high school he translated the *Antigone* of Sophocles. When his more fortunate friends went to college, Robinson studied at home and wrote

with a fierce intensity and loneliness. Finally, his need for treatment in Boston of an infected ear persuaded his family to let him enter Harvard (1891-3). Extremely shy and poor, he met the confident Robert Morss Lovett only when the latter called to reject a poem Robinson had submitted to the Harvard Monthly, which later condescended to print his "Richard Cory." His poems in the Harvard *Advocate* revealed a disciplined originality in marked contrast to a Keatsian poem by William Vaughn Moody, whom Robinson must have envied for his easy successes. By 1893, death and poverty in the Robinson family forced the young poet to return to Maine. Now his eyes were so poor that he had to restrict his reading, but he did have time—if not exactly leisure—to write. He chose poetry as a profession with a desperate faith in his destiny, and in his conviction that a poem should present the emotions of the human heart with a fearless simplicity, devoid of the cosmetics and conceits that were winning an early audience for other young poets of his day. But seldom has a faith been so sorely tried. He managed to find $52 to print 300 paper-bound copies of *The Torrent and the Night Before* (1896), which he mailed to critics and prominent literary people, including Hardy and Swinburne, who expressed their admiration for his work. The newspaper reviews were more favorable than the periodicals. He was praised by Professor Carpenter of Columbia and Edward Eggleston wondered that he had not heard of this poet before. But the *Bookman* regretted that the world was not beautiful for him. Robinson replied, in a letter which they printed: "The world is not a 'prison-house,' but a kind of spiritual kindergarten where bewildered infants are trying to spell God with the wrong blocks." He spent the following winter in New York, and in December, 1897, a small Boston house published *The Children of the Night*, which was barely noticed by reviewers, although it did lead to a few important friendships. Laura Richards had already introduced him to Hays Gardiner, a Harvard professor, who helped him to secure a position as clerk to President Eliot of Harvard. He acquired new friends, including Josephine Preston Peabody and George Pierce Baker (qq.v.), but he liked neither Eliot nor his job. He returned to New York and became a time checker in the construction of a subway. Mrs. Richards put up the money to have *Captain Craig* (1902) published, but the reviewers attacked Robinson for his intellectual irony, which they saw as obscurity. Meanwhile, he had not sold a poem in ten years, although the smaller journals occasionally printed his work. In Gardiner he had tried to write stories, but they too were rejected. After 9 months with the subway, he took an advertising job in Boston. But Henry Richards, Jr. had shown a copy of *The Children of the Night* to one of his students at Groton, who in turn sent the book to his father, Theodore Roosevelt. In March, 1905, Robinson received praise from the White House, and the President secured a position in the New York Customs House for the poet, at $2,000 a year. Robinson held the job until Taft be-

came President, and then resigned because he had looked upon it as a deliberate subsidy intended by Roosevelt to give him leisure for writing poems at his Wall Street desk. Roosevelt reviewed a reissue of *Children* for *Outlook*, and was much criticized by other reviewers for sticking his presidential nose into their business. From 1905 to 1913, Robinson's only sale to a magazine was a single poem, published in the *Atlantic* (1907). Scribner's published *The Town Down the River* in 1910, and for the first time the critics showed some sign of perceiving his quality, but the book sold poorly. Aided by friends, Robinson was able to devote his entire time to writing after 1910, and spent the summer months at the MacDowell colony at Petersboro, New Hampshire. With the publication of *The Man Against the Sky* (1915), he was finally recognized, 35 years after he had started writing poetry, as an important American man of letters. Aided by friends, alcohol, and great determination, he had survived endless disappointments and poverty, but he was too shy to take any advantage of fame when it finally arrived. His *Collected Poems* (1921) won the first Pulitzer Prize for poetry, and he won the award again in 1925 for *The Man Who Died Twice*, and in 1928 for *Tristram*, which sold 62,000 copies in a few months. Honorary degrees and awards were showered upon him, and when he died in 1935 he was recognized as the greatest American poet.

Like many of the intellectual poets who followed him, he was pessimistic concerning the threat of modern materialism, but he always maintained an optimism for the possibility of overcoming contemporary evils through human dignity and ideals. He maintained an ironic humor and a gentle love whether treating the garrulous *Captain Craig,* the misplaced "Miniver Cheevy," or the drunken Mr. Flood. Although he employed established metrical forms, he was far ahead of his time in his simplicity of statement and in his rejection of Victorian sentimentality. Like Browning, he frequently used the device of the dramatic monologue or dialogue, but in his own realistic manner and in the treatment of simple New England townsmen, New York failures, or historical heroes. (See also POETRY).

ROBINSON, Rowland Evans (1833 - 1900), a local-color writer of Vermont, displayed the Down East humor, dialect and philosophy in his sketches of rural life and sports. His work was published in a "Centennial Edition" (7 vols., 1933-6). Earlier works include *In New England Fields and Woods* (1896), and *Danvis Folks* (1894).

ROETHKE, Theodore (1908-   ), grew up in Saginaw, Michigan and was educated at the University of Michigan and Harvard. He has taught English at Lafayette College, Penn State, Bennington, and Washington. His poems have been collected in *Open House* (1944), *The Lost Son and Other Poems* (1948), and *Praise to the End!* (1951). He writes about ordinary life in ordinary language. His subject might be the pickle factory where he once worked or the scenes of a greenhouse. But his imagery and his coupling of ordinary words is extraordinary.

ROGERS, Will[iam] (1879-1935), Oklahoma cowboy and traveler with a Wild West show, incorporated a humorous monologue into his rope-trick act which later spread into a syndicated newspaper column. His wry sayings were in the homespun tradition of 19th-century newspaper humor. He was also a highly successful motion picture actor. His books include *The Illiterate Digest* (1924) and *Letters of a Self-Made Diplomat to His President* (1927).

RÖLVAAG, O[le] E[dvart] (1876-1931), born to a Norwegian family of fishermen, had little schooling before he was 14 when he was forced to devote full time to the fisheries. At 20 he came to America and worked on a South Dakota farm, where he managed to learn sufficient English and save enough money to begin college. In 1905 he graduated from St. Olaf College, Minnesota, and after a year of graduate study at the University of Oslo, Norway, returned to become a professor of Norwegian at the Minnesota school. In his *Letters from America* (1912), *Giants in the Earth* (1927), *Peder Victorious* (1929), and *Their Fathers' God* (1931), all written in Norwegian, he depicted the Norwegian immigrants and their pioneering adventures in Minnesota and the Dakotas, struggling against the harsh land with religious mysticism and fervor.

Romanticism, a critical term applied to that literature which emphasizes the personal, the subjective revelation of character, and the ideal in human behavior, usually expressed with imagination that is unrestrained by artistic dignity or decorum. Romanticism had replaced the neo-Classicism of the 18th century in England as American literature began to take shape, and our cultural situation only added impetus to its domination: the glorification of the primitive life of the frontier (Cooper, Thoreau); Transcendentalism (q.v.) (Emerson, Longfellow); democracy and the individual (Jefferson, Emerson, Whitman); and the sentimentality that was perhaps inevitable in a new country which had overthrown aristocratic notions of taste along with aristocratic government. It remains dominant in popular literature, but was replaced as a literary force by the various branches of Realism that emerged in the 20th century, pioneered by Stephen Crane Norris, and Dreiser.

ROSE, Aquila (c. 1695-1723), an English typographer, immigrated to Philadelphia about 1717. His neo-Classical *Poems on Several Occasions* (1740) were published by his son, Joseph Rose, and Franklin.

ROSENFELD, Paul (1890-1946), New York music critic and author,wrote a sensitive novel about *The Boy in the Sun* (1928), in addition to *Musical Portraits* (1920), *Musical Chronicle* (1923), *An Hour With American Music* (1929), and *Discoveries of a Music Critic* (1936).

ROURKE, Constance [Mayfield] (1885-1941), grew up in Ohio, studied at Vassar and the Sorbonne, and taught English at Vassar (1910-15). She made a study of the folk history of America which she reported in a fusion of criticism and history in such books as *Trumpets of Jubilee* (1927), *Troupers of the Gold Coast* (1928), *American Humor* (1931), *Davy Crockett* (1934), and *Roots of American Culture and Other Essays* (1942).

ROWLANDSON, Mary [White] (c. 1635-c. 1678), daughter of a wealthy proprietor in Lancaster, Massachusetts, and wife of a minister, was held for ransom by Indians for a three month period, which she described in *The Soveraignty & Goodness of God, Together with the Faithfulness of His Promises Displayed; Being a Narrative of the Captivity and Restauration of Mrs. Mary Rowlandson* (1682). This work is frequently reprinted in American literature texts as an example of 17th-century American prose at its best. It is also valuable for its concise descriptions of the times, the Indians, and the attitudes of the settlers.

ROWSON, Susanna [Haswell] (c. 1762-1824), author, educator, and actress, was brought to America at 5 by her family when her father, a naval lieutenant, was stationed in Massachusetts. After they returned to England in 1777, she published several novels, the most popular being *Charlotte Temple* (1791), often considered one of the early American novels since part of the action is set here. A sentimental and didactic romance, it was extremely popular in both America and England. She and her husband returned to the U. S. in 1793 as an acting team, and they played in several of her comedies, including *Americans in England* (1796). She continued to publish novels after abandoning the stage in 1797 to conduct a girls' boarding school near Boston.

RUKEYSER, Muriel (1913-), New York poet who uses an amalgamation of contemporary verse fashions to express indignation over social injustices. Her message is frequently obscured by her expression, so that it reaches only the perceptive, who do not need it. As an emotional expression of social indignation, however, her work achieves an intensity of feeling in the poems collected as *Theory of Flight* (1935), *A Turning Wind* (1940), *Wake Island* (1942), and *The Beast in View* (1944). Some of her later work included in *Selected Poems* (1951) makes use of classical themes to achieve passionate apprehensions of human dignity in a highly polished style. She is presently living in California.

RUNYON, [Alfred] Damon (1884-1946), sports writer who achieved considerable fame with his short stories concerning athletic and underworld characters. He was one of the first writers to exploit 20th-century American slang for its color, humor, and reality. *Guys and Dolls* (1932) was used as the basis for a musical comedy. His other collections of stories include *Take it Easy* (1938), *My Wife Ethel* (1940), *Runyon à la Carte* (1944), and *Short Tales* (1946).

RUSSELL, Irwin (1853-79), Mississippi lawyer and poet, was credited by Joel Chandler Harris with giving the first accurate portrait of the southern Negro. He wrote many of his poems in the Negro dialect. Well read in the classic poets, he might have had a distinguished literary career if he had not died so young. Harris collected his work for publication as *Poems by Irwin Russell* (1888), and the collection was enlarged for republication as *Christmas-Night in the Quarters* (1917).

S

ST. JOHN, Hector, pseudonym of Michel-Guillaume Jean de Crèvecœur (q.v.).

SALTUS, Edgar [Evertson] (1855-1921), New York philosopher who popularized Schopenhauer in *The Philosophy of Disenchantment* (1885), and wrote a long series of romantic novels of hedonistic disenchantment that included: *The Pace That Kills* (1889), *The Pomps of Satan* (1904), *Vanity Square* (1906), and *The Monster* (1912). He also translated from the French, wrote a biography of Balzac, published two volumes of short stories, and wrote histories of Biblical and other historical characters.

SANDBURG, Carl [August] (1878-), grew up in Galesburg, Illinois, the son of Swedish immigrants. His father worked on a railroad construction crew, and he took odd jobs from an early age and later traveled about the West, working in the wheat fields and as a dishwasher in hotels. He served in Puerto Rico during the Spanish-American War, and upon his return, worked his way through Lombard College where he was captain of the basketball team, and editor of the college paper. After graduating in 1902 he tried business and politics, and was secretary to the Socialist mayor of Milwaukee (1910-12). Meanwhile he had been writing poems, and had privately published a small volume of them in 1904. While he was a journalist in Chicago, *Poetry* Magazine published several of his poems (1914), and their original expression in colloquial, unadorned, free verse attracted considerable attention as well as controversy. In contrast to many experimental writers, Sandburg could be readily understood, and thus readily attacked or praised. *Chicago Poems* (1916) contained his celebrated "Fog," as well as "Grass," and was awarded the Helen Hare Levinson Prize of $200. In 1918 he visited Norway and Sweden as correspondent for the Newspaper Enterprise Association, and upon his return, became an editorial writer for the Chicago *Daily News*. In 1933 he retired to Harbert, Michigan, on the shore of Lake Michigan, to write and to raise prize goats. During the forties he moved to Chimney Rock, North Carolina. *Cornhuskers* (1918) won a special Pulitzer award, and was followed by *Smoke and Steel* (1920), *Slabs of the Sunburnt West* (1922), *Selected Poems* (1926), and *Good Morning, America* (1928).

In these works he displays the common American idiom and thought, and a sympathy for the common people, that is reminiscent of Whitman; yet, he achieves a concise realism and concrete specification that is distinctly his own. His interest in American folklore is displayed in his collection of folk songs, *The American Songbag* (1927). This interest was also embodied in *The People, Yes* (1936), which expresses his confidence in America and in her people through a panorama of folk history. He devoted a number of years to research, planning, and writing of his great biography of Lincoln, published as *Abraham Lincoln: The Prairie Years* (2 vols., 1926), and *The War Years* (4 vols., 1939), which received the Pulitzer Prize. The 6 volumes were published together in 1954 as *Abraham Lincoln*.

Always the Young Strangers (1953) is a prose autobiography. Mr. Sandburg is an accomplished guitarist, and recordings have been made of his singing ballads in a folksy, resonant voice while accompanying himself on the guitar. He was honored on his 75th birthday (1953) by a televised testimonial banquet and special awards from The Poetry Society of America. In an age when serious poetry has tended to retreat from the sympathies and understanding of the common people, Sandburg has remained as a spokesman of, by, and for the people, without sacrificing his artistic integrity.

SANDOZ, Mari (1900-), Nebraska novelist, has taught creative writing at the University of Wisconsin, and now lives in New York, frequently spending her summers in the West. Her novel, *Old Jules* (1935), about the pioneer farming adventures of her Swiss emigrant father, received both critical and popular acclaim. Her other novels include *Slogum House* (1937); *The Tom-Walker* (1947); *Crazy Horse* (1942), a biography of the Sioux Indian chief, and a history of *The Buffalo Hunters* (1954).

SANTAYANA, George (1863-1952), was born in Madrid of Spanish parents, but was brought by his mother to Boston when he was 9. He was educated at Harvard, receiving his Ph. D. in 1889, when he became a professor of philosophy at that institution. He also studied in Berlin and at Cambridge, and lectured at Oxford and the Sorbonne. In 1912 he received an inheritance that allowed him to retire to Oxford, later to Paris, and eventually to Rome, where he lived quietly for the remainder of his life. He published *Sonnets and Other Verses* in 1894; a verse play, *A Theological Tragedy* in 1899 (revised, 1924); *A Hermit of Carmel* in 1901, and his collected *Poems* in 1923. His one novel was *The Last Puritan* (1935), which was a Book-of-the-Month Club selection. His other publications have been in the discipline of philosophy, including studies of *The Life of Reason* (5 vols., 1905-6); *The Sense of Beauty* (1896); and *Philosophical Opinion in America* (1918). *Persons and Places* (1944) presents his memoirs, which are continued in *The Middle Span* (1945).

SARETT, Lew (1888-), Rocky Mountain

poet and professor of English at Northwestern University, wrote poetry concerning the history of the western frontier, published as *Many Many Moons* (1920), *The Box of God* (1922), *Slow Smoke* (1925), *Wings Against the Moon* (1931), and *Collected Poems* (1941).

SARGENT, Epes (1813-80), journalist, editor, poet and dramatist, wrote romantic novels, including *Fleetwood; or, the Stain of Birth* (1845); the plays, *Velasco* (1837) and *The Priestess* (1854); and *Songs of the Sea with Other Poems* (1847). He was a member of a distinguished Boston family, and was respected for his editorial accomplishments in New York.

SAROYAN, William (1908-   ), the son of Armenian immigrants to California, began working at eight and left school at fifteen, but studied literature in his spare time and made independent and original decisions concerning the type of writer he wanted to become. In 1934 *Story* magazine (q.v.), having made original decisions concerning the type of writer it wanted to publish, discovered Saroyan, and brought out his *The Daring Young Man on the Flying Trapeze,* a collection of his stories about the impression left on him by people, incidents, or perceptions. Other collections followed, including *Inhale and Exhale* (1936); *Little Children* (1937); *The Trouble with Tigers* (1938); and *Love, Here Is My Hat* (1938). *My Name is Aram* (1940) is an exquisitely gentle story of a growing boy. Saroyan's love for people has sometimes provoked the

criticism that he is sentimental; however, his very human sense of humor and the quality of poetry he achieves in revealing the dignity of vulgar persons protects him from sentimentality except on the few occasions when he becomes didactic about ideas. *The Human Comedy* (1943) is a novel based on his motion picture scenario of that title. *The Time of Your Life* (1939), a Pulitzer Prize play, was as original as his stories, presenting a series of people and incidents in a San Francisco waterfront bar; and *The Beautiful People* (1941) followed the same pattern. Saroyan writes easily and quickly, and his success derives from his unique talent and personality rather than from studied construction or literary discipline.

Saturday Review of Literature, The (1924-   ), a weekly journal containing book and drama reviews and fairly conventional literary commentary. The editors have been H. S. Canby (1924-36), Bernard DeVoto (1936-8), and Norman Cousins (1940-   ).

SAWYER, Lemuel (1777-1852), native North Carolinian and member of Congress for eight sessions, wrote a remarkable comedy *Blackbeard* (1824) about contemporary Southern politicians, treasure-hunters, and backwoodsmen. It is rich in dialect. Other major works are *The Wreck of Honor* (1824), a tragedy; *A Biography of John Randolph of Roanoke* (1844), a scathing attack on the noted man written under the guise of friendship; and *Autobiography of Lemuel Sawyer* (1844), a shameless exposé of a life of frippery and extravagance.

SAXE, John Godfrey (1816-87), son of a prosperous Vermont mill owner, was active in politics and journalism, but won his reputation as a witty after dinner speaker and as the composer of light verse, modeled on Holmes, but never equalling him. He published ten volumes of light verse which was very popular in his day, but is now mainly forgotten.

SCHORER, Mark (1908-   ), grew up in Wisconsin and was educated at the University of Wisconsin and at Harvard. He has taught at Wisconsin, Dartmouth, Radcliffe, and Harvard, and is now a professor at the University of California. He has written three novels, *A House Too Old* (1935), *The Hermit Place* (1941), and *Wars of Love* (1954). In addition to critical articles he has written a monumental definition of *William Blake: The Politics of Vision* (1946). Many of his short stories have appeared in the O. Henry and O'Brien collections, and are included in *The State of Mind*.

SCHULBERG, Budd (1914-   ), son of a film producer, grew up in Hollywood, attended Dartmouth College, and has taught creative writing at Columbia University. His first novel, *What Makes Sammy Run?* (1941), is a satirical study of an opportunist who achieves Hollywood fame after a career of cheating and grasping. *The Harder They Fall* (1947) is an exposé of professional boxing, and *The Disenchanted* (1950), said to be based on the life of Fitzgerald, presents a highly successful novelist of the twenties later meeting tragedy and reality.

SCHWARTZ, Delmore (1913-   ), New York poet and critic, graduated from New York University and has been an instructor at Harvard. His books of poems include *In Dreams Begin Responsibilities* (1938); a verse play, *Shenandoah* (1941); and *Genesis* (1943). He is literary in derivation and modernistic in style.

SCOTT, Evelyn (1893-   ), grew up in New Orleans, where she wrote and published stories at the age of 14. At 20 she moved to Brazil for three years which she described in *Escapade* (1923). Her antipathy to middle-class morality was demonstrated in the novels, *The Narrow House* (1921), and *Narcissus* (1922). Her more popular novels are *The Wave* (1929), a Civil War story; and *Blue Rum* (1930, published under pseudonym of E. Souza), set in Portugal. *Bread and a Sword* (1937) expresses her mature liberalism in dealing with the problems of the author in a contemporary society. She has also published two volumes of poetry, *Precipitations* (1920), and *The Winter Alone* (1930), as well as a collection of stories, *Ideals* (1927).

Scribner's Magazine (1887-1939) was founded by Charles Scribner the younger (1854-1930) after his father had sold *Scribner's Monthly*, which became *The Century* (1881-1939). The new *Scribner's* became distinguished for its publication of Henry and William James, Kipling, Harte, Cable, Edith Wharton, and Stephen Crane. In this century it was the first literary journal to publish Hemingway and Thomas Wolfe.

SEABURY, Samuel (1729-96), Connecticut minister who wrote skilfully-presented arguments against the American revolutionary cause under the pseudonym of "A Westchester Farmer," which were collected as *Discourses on Several Subjects* (3 vols., 1791-8). Such pamphlets as "The Congress Canvassed" and "A View of the Controversy between Great Britain and Her Colonies" (1774), were answered by Hamilton.

SEAGER, Allan (1906-   ), grew up in Michigan and Tennessee, graduated from the University of Michigan, and took his M.A. degree at Oxford as a Rhodes scholar. He served as an editor of *Vanity Fair* (1934), and has taught literature and creative writing at Bennington College and at the University of Michigan. His stories, which appeared in the O'Brien, O. Henry, and Foley collections, were published as *The Old Man of the Mountain* (1950). His first novel, *Equinox* (1943), was a tragedy of a former foreign correspondent baffled by the unnatural love of his daughter. *The Inheritance* (1948) concerns a young man dominated by the memory of his father.

SEALSFIELD, Charles (1793-1864), the adopted name of Karl Postl after his escape from a Prague monastery to Switzerland and later to the United States. After traveling in the Southwest he wrote several books about frontier life, which were translated from German into English as *Tokeah; or, The White Rose* (1928); *The Cabin Book; or, Sketches of Life in Texas* (1844); and *Frontier Life*

(1856).His collected works were published in 15 volumes in 1945.

SEAMAN, Elizabeth Cochrane; See "Bly, Nelly."

SEDGWICK, Anne Douglas (1873-1935), English novelist in the manner of Henry James, was born in New Jersey, but resided in England from the age of nine. Her novels include *Tante* (1911) and *The Encounter* (1914), both studies in genius; *Adrienne Toner* (1922); *The Little French Girl* (1924); and *Dark Hester* (1929).

SEDGWICK, Catharine Maria (1789-1867), a wealthy Massachusetts philanthropist, began writing romantic novels to arouse social conscience in behalf of the humbler virtues, and thereby provided a valuable picture of 19th-century customs. Her novels include *A New-England Tale* (1822), *Redwood* (1824), and *Married or Single?* (1857).

SEEGER, Alan (1888-1916), New York poet and Harvard graduate, was killed in the First World War while fighting with the French Foreign Legion. His poem, "I Have a Rendezvous with Death," published in the *North American Review* (Oct., 1916), became extremely popular and was widely reprinted. His *Poems* were published in 1916, and his *Letters and Diary* in 1917.

SELDES, [Vivian] Gilbert (1893-   ), journalist and drama critic, made a critical study of *The Seven Lively Arts* (1924)), which examined comic strips, movies, and songs. He has also written several studies of American civilization, a novel—*The Wings of the Eagle* (1929), and detective stories (under the pseudonym of Foster Johns).

SERVICE, Robert W[illiam] (1874-   ), English-born Canadian, wrote heroic verse about his experiences in the Yukon which were extremely popular, and were published in *The Spell of the Yukon* (1907). He also wrote adventure novels, and his reminiscences were published as *Ploughman of the Moon* (1945).

SEWALL, Samuel (1652 - 1730), was born in England of colonial parents who returned to Boston when he was nine. He was graduated from Harvard in 1671, and tutored there for many years before he became active in politics. One of the judges at the Salem witchcraft trials, he later publicly confessed his guilt in damning 19 innocent persons. He is best remembered for his *Diary*, published by the Massachusetts Historical Society (3 vols., 1878-82), an intimate account of the manners and attitudes in New England from 1674 to 1729. In *The Selling of Joseph* (1700) he pleaded for the rights of slaves, and in his letters he advocated humane treatment for the Indians. Wise and tolerant, yet conventional and narrow in many of his views, he is himself a picture of 17th-century New England, and his self portrait in the *Diary* makes it the more valuable as a source book.

Sewanee Review (1892-   ), a literary quarterly published by the University of the South in Sewanee, Tennessee, which in 1944 turned from the normal academic patterns to an emphasis on modern literature under the editorship of Allen Tate.

SHAPIRO, Karl [Jay] (1913-   ), Baltimore poet whose *V-Letter, and Other Poems* (1944) won the Pulitzer Prize. His other works include *Essay on Rime* (1945), a criticism of modern poetry written in verse, and *Trial of a Poet* (1948). He is editor of *Poetry* Magazine (q.v.).

SHAW, Henry Wheeler, See "Billings, Josh."

SHAW, Irwin (1914-   ), Brooklyn-born, and educated at Brooklyn College, wrote dramatizations of the comic strips for the radio before his play, *Bury the Dead* (1936), provided an entrance to Hollywood, where he wrote scenarios. His stories, written in the "hard-boiled" school of narrative, have appeared in the O'Brien collections, and in 1944 he received the O. Henry Memorial Award. They have been collected as *Sailor Off the Bremen* (1939), *Welcome to the City* (1942), and *Act of Faith* (1946). His novel, *The Young Lions* (1948), pictures Semitism and anti-Semitism in a Bavarian concentration camp.

SHEEAN, [James] Vincent (1899-   ), after studying at the University of Chicago for three years, became a European correspondent for the Chicago *Tribune* (1922-5), and later extended his coverage to the Far East. He records his reaction to his first-hand view of history, which included the Fascist march on Rome, in *Personal History* (1935). This was followed by publication of further studies of contemporary European history: *Between the Thunder and the Sun* (1943), *This House against This House* (1946), and *Not Peace but a Sword* (1939). He has also written several novels, including *Sanfelice* (1936), and *A Day of Battle* (1938).

SHELDON, Edward [Brewster] (1886-1946), was born in Chicago, and took his B.A. and M.A. degrees at Harvard, where he studied under George Pierce Baker (q.v.). His first play, *Salvation Nell* (1908), was a success, and was followed by *The Nigger* (1909) and *The Boss* (1911), both concerning politics. *The High Road* (1912) pictured a farm woman driving her husband to candidacy for President. *Romance* (1913) played for several years, and was filmed in 1930 with Greta Garbo. *Dishonored Lady* (1930) starred Katharine Cornell. He wrote *Lulu Belle* (1926) with his nephew, Charles MacArthur, and *Bewitched* (1924) with Sidney Howard.

SHELLABARGER, Samuel (1888-1954), took his B.A. at Princeton and his Ph.D. at Harvard (1917), and after service in World War I became an assistant professor of English at Princeton. He later resigned and settled in Switzerland, and still later lived in England and France, before returning to America. In addition to detective stories and romances (published under the pseudonyms of "John Esteven" and "Peter Loring") he wrote

scholarly biographies of *The Chevalier Bayard* (1928) and *Lord Chesterfield* (1935). His historical novels have been extremely popular. *Captain from Castile* (1945) deals with Spanish conquests in the 16th century. *Prince of Foxes* (1947) is concerned with the plottings of Cesare Borgia. *The King's Cavalier* (1950) is set in France at the time of the Bourbon conspiracy against Francis I.

SHERMAN, Frank Dempster (1860-1916), a Columbia professor of architecture, was famous for his wit and charm both in and out of his verses, published in *Madrigals and Catches* (1887), *Lyrics for a Lute*, and other volumes, and collected in *Poems* (1917). He also wrote delightful verse for children under the pseudonym of Felix Carmen.

SHERWOOD, Robert [Emmet] (1896 -), began his career in New Rochelle, New York as editor of the *Children's Life*, but in 1904 gave up that position to devote himself to the rewriting of *A Tale of Two Cities*, feeling that he could improve on Dickens. Later in his life he attended Milton Academy and Harvard, then fought with the Canadian Royal Highlanders and was gassed at Arras and wounded at Amiens. In 1919 he again picked up his editorial career, serving as movie and drama critic for *Vanity Fair*, *Life*, and the New York *Herald*. With the success of his first play, *The Road to Rome* (1927), he gave up journalism to become a professional dramatist and scenarist. He wrote 4 additional successes before *Reunion in Vienna* (1931) made him famous, starring

Alfred Lunt and Lynn Fontanne as the exiled Hapsburgs in a nostalgic comedy. *The Petrified Forest* (1935) examined human and political values against a background of the symbolic National Monument in Arizona. *Tovarich* (1936) was a successful adaptation from the French play of Jacques Deval. He won the Pulitzer Prize for *Idiot's Delight* (1936), an anti-war play that intelligently saw the Second World War brewing. He was again awarded the Pulitzer Prize for *Abe Lincoln in Illinois* (1938), and for *There Shall Be No Night* (1940). During the war he had prophesied, he worked closely with President Roosevelt, and reported this activity in *Roosevelt and Hopkins* (1948). Sherwood writes his plays quickly, but he plans them in great detail before he begins the actual composition.

SHILLABER, Benjamin Penhallow (1814-90), began his career as a printer in Portsmouth, N. H., and later became a journalist in Boston, where he won sudden fame for his creation of "Mrs. Partington" for the Boston *Post* in 1847. Ruth Partington employed terrible malapropisms as she discoursed congenially on every conceivable topic, from gardening to Calvinism, in blissful ignorance. She is considered to be Mark Twain's model for Tom Sawyer's Aunt Polly. In 1851 Shillaber set up a comic weekly, *Carpet Bag*, which published Twain's first work, "The Dandy Frightening the Squatter" (May 1, 1852), signed "S. L. C." Shillaber's sketches were collected as *Life and Sayings of Mrs. Partington* (1854), *Mrs. Partington's Knitting*

Work (1859), *Partingtonian Patchwork* (1873) and *Mrs. Partington's Grab Bag* (1893).

SHORT, Bob, pseudonym of A. B. Longstreet (q.v.).

SHORT STORY IN AMERICA:
   The short story, in its modern form, was born and developed in America. There had been tales and stories, of course, long before Edgar Allan Poe set down his requirements for the short story. The stories told by Boccaccio's exiles from the plague; the Latin tales in the Gesta Romanorum; and the short prose narratives of the ancient Egyptians, Hebrews, Greeks and Arabs are all, in a sense, short stories. Other narratives, the tales told by Chaucer's pilgrims to Canterbury and the lays of Marie de France, are very similar to the short story except for the medieval convention of presenting such narrative in verse. The "Lay of the Divided Blanket" has a realistic surprise ending very similar to the stories of O. Henry. But even in early 19th-century America, short narratives were called "tales." It was in a review of Hawthorne's *Twice-Told Tales* that Poe first set down his critical perceptions as to what a short story should be (1842). He felt that it should be sufficiently short to permit the reader to finish the work in a single sitting; for any break in the reading would disturb the dramatic effect. He advised the author to conceive, initially, of a single effect which he would intend to produce, then invent incidents, characters, and situations to accomplish this inten-

tion. The structure should, ideally, be so economical and intense a realization of the single effect that not a line could be omitted without considerably weakening the power of the story. He himself specialized in the Gothic story of terror, where the single effect could be extremely powerful.

With the emergence of the local color tradition, such writers as Bret Harte, writing about California; Kate Chopin, describing Creole Louisiana; and Charles Egbert Craddock (Mary Noailles Murfree), using the Tennessee mountains as a background, developed realism in using the story form. Ambrose Bierce, meanwhile, continued in the development of the horror story. Henry James came closer to the structural unity of Poe than any other story teller of the later 19th century, especially in "The Turn of the Screw." But James found difficulty in following the Poe restriction on length.

At the turn of the 20th century, O. Henry (William Sydney Porter) developed a type of story which was to influence other writers for years to come. Like Harte, he used a conversational, anecdotal style, but his major contribution was the surprise or "snap" ending. In later years his formula was used particularly by the writers for the popular magazines of wide circulation, which came to be called "slick" (*Colliers, Saturday Evening Post,* etc.), because they were published on a smooth paper. The poorer quality of "action" story magazines came to be designated as the "pulps," and such magazines as *Atlantic Monthly* and *Harper's* were called "quality."

With the new interest in realism which was prevalent early in this century, and under the influence of Maupassant and Chekhov, a new type of story appeared which substituted representationalism for structure, and importance of material or message for importance of artistic effect. Sherwood Anderson was the master of this form, and had a wide influence on future writers. Apprehension of experience, and some revelation of truth, either poetic or didactic, was the object. Meanwhile, a new school of story criticism developed to oppose the O. Henry influence. Edward J. O'Brien, editor of the annual anthologies of the *Best Short Stories* (1921-), emphasized, in his selections, the new approach, and later he was joined by Whit Burnett and Martha Foley, editors of *Story* magazine, who rarely accepted a story with a surprise ending. Their influence on young writers was especially strong because, for a time, they offered one of the few paying markets for experimental short stories in which the unknown writer felt that his work would receive as much attention as the established or "name" author. Their rate of payment was $25, whether the author was Lord Dunsany or a Western Union messenger boy named William Saroyan. During the thirties, almost every young short story writer of serious pretensions aspired to place his work in *Story* magazine.

In addition to the O'Brien anthology of prize stories, a rival anthology of the O. Henry Memorial Award stories was also published annually, using the Porter style and technique as a criterion of quality. Thus, the tenets of the O'Brien followers were not permitted to dogmatize the American short story, and recognition was provided for the excellent stories which sometimes appear in the commercial magazines. It is interesting to note that most major short story writers of America have appeared in both anthologies.

O'Brien and his followers were not, however, opposed simply to the O. Henry type of story; they objected principally to the rabid commercialization of the short narrative, which they identified with the O. Henry story since his technique and style could best be exploited for profit in the growing magazine trade. Douglas Bement noted in 1931 that "the news-stands are flooded with hundreds of magazines ranging from intimate boudoir confessions to profound psychological studies; at least three of these periodicals which specialize in short stories have a combined circulation of over six million copies a week, a fact which makes them, in the opinion of astute publicists, the best media in the world for selling tooth-brushes, automobiles, and canned soup" (Introduction, *Weaving the Short Story* by Douglas Bement, 1931). Bement estimated that besides the great bulk of short fiction published weekly, 500 stories were rejected by the magazines for every one published. American democracy, with its emphasis on universal education and its concept that anyone could be anything he desired, had promoted a great interest in writing. It was considered glamorous to be a writer, and every state had authors' organizations, the usual requirement for membership being that the applicant must have

earned a certain sum of money by his pen. Trade journals for writers advertised innumerable correspondence courses, each one promising the magic formula for commercial success. It was and still is paradoxical that these professional journals are aimed at the non-professional writer, and their message has been consistent: "you have not yet broken through the editorial barrier, but you can do it if you stick with us." The reward offered is financial success. The story would be the one narrative form which would most quickly appeal to the large mass of aspirant writers. Since most of them had little or no conception of literature as an art, and confined their reading to stories of entertainment, they were most familiar with the "slick" magazine story. Most of them were impatient for glory and riches, and the novel would take much too long to write. Most pulp and slick stories were written according to formula—get your hero in a predicament, throw rocks at him, then have him fight his way out; therefore, the aspirants had some hope of achieving success, for they simply had to learn the formula. The formula alone was enough for the pulp market, but the slick magazines required narrative ability in its presentation.

Even though some excellent stories were written within this formula and for the commercial magazines, most of the creative and experimental work was done for academic quarterlies, and for *Story* Magazine, which accepted no advertising. The so-called "quality" commercial magazines printed excellent stories written in more conservative traditions.

Meanwhile a new and quite different commercial market appeared with the founding of *The New Yorker*. Although its requirements were perhaps stereotyped, it presented a new type of story intended for intelligent and sophisticated subscribers. Less sophisticated readers complained that when they had read a *New Yorker* story they often turned the page, expecting more. Actually, these stories, at their best, did and do have a beginning, middle, and end, but they are not blatantly announced. (The modern stories which do not present the three Aristotelian parts are more properly called "sketches.") Another characteristic of the *New Yorker* story is its stylistic resemblance to the novel: the pace is leisurely, and contrary to classical precepts, characters are not merely presented but are sometimes developed.

Another unusual market was *Esquire* magazine, which published many of the finest stories of Hemingway, Steinbeck, Saroyan, and other writers of the first rank, as well as beginning writers. *Esquire* was directed toward male readers of the upper middle-class, and the curt, masculine style of Hemingway dominated most of its stories.

From the academic point of view, the story has declined as a literary study, and it is nearly impossible for a writer to make a national reputation purely in this form. A few of the exceptions are Sally Benson, Mark Schorer, and Allan Seager, but each has strengthened his reputation in other fields. James Thurber, S. J. Perelman and Peter DeVries have succeeded remarkably well with the humorous story of literary quality. Reputations tend to become based on books rather than upon magazine publication, and the market for stories in collection is not large except in the field of humor.

SIGOURNEY, Lydia Huntley (1791-1865), of Norwich, Connecticut, was a popular writer of 67 published books, and earned a substantial income, after the failure of her husband's hardware business, writing sentimental and pious verses for the popular magazines.

SILL, Edward Rowland (1841-87), orphaned at 12, was reared by an Ohio uncle, graduated from Yale, and sailed around the Horn to California, where he worked at various jobs. Returning East, he studied at the Harvard Divinity School, which prepared him merely for a life of agnosticism. He became a teacher in New York and Ohio, and a professor of English at the University of California. He might have become the Matthew Arnold of America had his health permitted a more serious concentration on his verse, for his poetry had a simplicity and spontaneity unusual in his time. He spent his last years in Ohio, writing essays (collected as *Prose*, 1900) and poems for the magazines. *The Venus of Milo and Other Poems* was privately printed in 1883, and his collected *Poems* were issued in 1902.

SIMMS, William Gilmore (1806-70), Charleston poet and novelist, was apprenticed to a druggist as a child, and later studied law. When he was 21 he published two volumes of sentimental verse. During a three-year residence in New York he wrote a psychological novel, *Martin Faber* (1833), and *Guy Rivers* (1834), the latter establishing the pattern of his later work in the romanticizing of the southern frontier. *The Yemassee* (1835), perhaps his best novel, is a historical melodrama of Indian warfare in South Carolina, and *The Partisan* (1835) concerns the Revolution. He was financially ruined by the Civil War, and although he continued to turn out hack work, his powers declined as his need for money increased. Called "the Cooper of the South," and serving as a self-appointed publicity agent for southern ideals, he was never completely appreciated by a southern aristocracy which could not overlook his humble origins. He wrote rapidly, and perhaps tried too hard to do too much. He was continually split between his loyalties and southern realities, and his melodramatic qualities conflicted with his realistic perceptions. His other major works include *Richard Hurdis* (1838), *Border Beagles* (1840), *Beauchampe* (1842), *Charlemont* (1856), and *The Wigwam and the Cabin* (1845).

SINCLAIR, Upton [Beall] (1878-   ), born to a prominent but destitute Baltimore family, has supported himself by writing since the age of 15. While studying law at City College, New York, he wrote action stories and dime novels. While studying music at Columbia he wrote 50 stories about West Point for Street and Smith, as well as another series about Annapolis. In 1900 he settled in the country to write pot-boilers to support his family while trying more serious novels that failed to sell, and in this period he became increasingly active as a Socialist. In 1904 his fortunes were miraculously changed by an assignment to investigate stockyard conditions in Chicago for a Socialist paper. His exposé, published as *The Jungle* (1906), was responsible for the Pure Food and Drug act, and was a best seller in 17 languages. He used the profits to found Helicon Home, a Socialist colony in New Jersey, where young Sinclair Lewis worked for a time as janitor. Later Sinclair traveled in Europe, and in 1915 settled in California where he has been active in politics, even running for governor. He has continued to turn out scores of books, stories, and pamphlets at the potboiler pace, supporting strikers, of the Colorado coal mines in *The Brass Check* (1919); exposing higher education in *The Goose-step* (1923), and lower education in *The Goslings* (1924); studying the film industry in *Upton Sinclair Presents William Fox* (1933), and the automobile industry in *The Flivver King* (1937). In 1940, with *World's End*, he began his series of novels about Lanny Budd, illegitimate son of a munitions manufacturer, who is a *Presidential Agent* (1944) for Franklin D. Roosevelt, and meets all of the celebrated and notorious characters of contemporary history. These books are written in a narrative style that combines Street and Smith heroics with a sociological investigation of the modern world.

Smart Set, The (1890-1930), originally a New York society magazine, became a literary journal at the turn of the century, and published O. Henry's first short story, and the work of Cabell, LeGallienne, Nathan, and Mencken. Willard Huntington Wright took the magazine still further afield during his brief editorship (1913-14), publishing fresh American material as well as the work of George Moore, Joyce, D'Annunzio, D. H. Lawrence, and Ford Madox Ford. Mencken and Nathan, who followed Wright as the editors, continued his policies, publishing the early contributions of O' Neill, Fitzgerald, Krutch, Mumford, and Frank. Before the magazine was purchased by Hearst (1924) and assumed a more conventional pattern, it was one of the most important literary forces in the country.

SMITH, Betty [Wehner] (1904-), grew up in Brooklyn, New York, leaving school to go to work after the eighth grade. After her marriage, she studied at the University of Michigan, winning the Avery Hopwood drama award, and later at Yale with George Pierce Baker. In 1947 she became a drama consultant and lecturer at the University of North Carolina. She has published 70 one-act plays and edited two drama books. She first won national prominence with the publication of *A Tree Grows in Brooklyn* (1943), a realistic portrait of a girl growing up in the Brooklyn slums. Her second novel portrays the vain hope of a poor young couple in Brooklyn that *Tomorrow Will Be Better* (1948).

SMITH, Charles Henry, See "Bill Arp."

SMITH, Elizabeth Oakes (1806-93), the wife of Seba Smith, wrote popular novels in the sentimental tradition, usually under the pseudonym of Ernest Helfenstein, including *Bald Eagle* (1867), and *A Legend for Christmas* (1848).

SMITH, Francis Hopkinson (1838-1915), Baltimore engineer, wrote charming travel sketches as well as novels, including a local-color presentation of *Colonel Carter of Cartersville* (1891), who had come upon hard times; *Tom Grogan*, presenting a female Irish contracting stevedore; and *Kennedy Square* (1911).

SMITH, John (1580-1631), became a soldier of fortune at 16, and had a variety of adventures on the continent, including imprisonment in Constantinople, before he joined the Virginia Company at 26 in the founding of Jamestown (1607). Although he was, apparently, the bravest and most capable officer of the company, and saved the others by his ability to deal with the Indians, his enemies had him discredited in England. He also explored New England for a group of London merchants who were disappointed that he brought back no gold. He settled in London to write of his adventures in a clear and forceful style, with sharp observations concerning the new country. His complete works were edited and published by Edward Arber in 1884, and a second edition was published in 1910.

SMITH, Lillian [Eugenia] (1897-), grew up in Florida, and studied at Piedmont College, the Peabody Conservatory of Music in Baltimore, and Columbia University. She taught in Huchow, China for three years, then returned to Georgia to direct a girls' camp and edit the magazine, *The South Today*. Her novel, *Strange Fruit* (1944), pictures the strife resulting from the love of an educated Negro girl for a white man, culminating in murder and lynching. The book sold more than three million copies and won her awards for its contribution to better race relations.

SMITH, Richard Penn (1799-1854), wrote romantic melodramas, most of them taken from the French. Among his original works were *William Penn, or The Elm Tree* (1829), *The Triumph of Plattsburg* (1830), and *Caius Marius* (1831), which has been lost. *The Eighth of January* (1829) used a French pattern for a dramatization of Jackson's victory at the Battle of New Orleans. He also wrote the novel, *The Forsaken* (1831), a story of the Revolution, and is considered to be the author of *Col. Crockett's Exploits and Adventures in Texas* (1836).

SMITH, Seba (1792-1868), Maine newspaper publisher who achieved fame for the letters he published in his *Portland Courier* "by Major Jack Downing of Downingville," which were widely reprinted and imitated, and eventually created a tradition of newspaper humor that extends to Will Rogers in our own time. Despite his military title, Major Downing wrote in a rustic Yankee dialect with homely sagacity, or often with a disarming simplicity that disguised

Smith's shrewd observations on local and national politics. The Jackson administration was satirized through the device of making Downing the President's confidant. A pirated edition of the letters appeared before Smith could bring out his own *The Life and Writings of Major Jack Downing of Downingville* (1833). After suffering business reverses, Smith settled in New York where he held numerous editorial positions and continued writing both serious and satirical work, including: *Way Down East* (1854), a collection of realistic portraitures of New England life; *New Elements of Geometry* (1850); and a satire on T. H. Benton's *Thirty Years' View of the American Government* which Smith called, *My Thirty Years Out of the Senate* (1859).

SMITH, Thorne (1892-1934), first won success with a novel concerning his Navy experiences during the First World War, *Biltmore Oswald, The Diary of a Hapless Recruit* (1918). However, he did not develop his famous style of ribald humor until the publication of *Topper* (1926), the story of two comic ghosts haunting a staid banker. *The Night Life of the Gods* (1931), another fantasy, brings statues to life. *The Bishop's Jaegers* (1932) captures a rare assortment of characters, including the Bishop, on a lost ferry-boat that comes aground at a nudist colony, affording Smith the opportunity to satirize conventional mores through the completely opposite mores of the colony, where a young woman who appears clothed is considered to be up to no good.

SMITH, Winchell (1871-1933), New York actor, producer and playwright, first introduced the plays of Shaw to New York audiences in 1904. With Frank Bacon he wrote *Lightnin'* (1918), a sentimental comedy which achieved a phenomenal success. He also dramatized *Brewster's Millions* (1916), the novel by G. B. McCutcheon (q. v.).

SNOW, [Charles] Wilbert (1884-), born on a Maine lighthouse station, was a deep-sea fisherman before studying at Bowdoin and Columbia for a career as a professor of English. His direct and honest poems, written in the Wordsworth tradition and using the virile language of common men, have been collected in *Maine Coast* (1923), *The Inner Harbor* (1926), *Down East* (1932), *Selected Poems* (1936), *Before the Wind* (1938), and *Main Tides* (1940).

Southern Agrarians: a term applied to a group of Southern writers who published *The Fugitive* (1922-5) in Nashville and the *Southern Review* (1935-42) at Baton Rouge, Louisiana. The members included J. C. Ransom, J. G. Fletcher, Allen Tate, R. P. Warren, and Donald Davidson. Politically, they advocated an agrarian economy for the South, and in literature they were founders of the New Criticism (q.v.) in America.

Southern Review, The (1935-42), was published at Baton Rouge, Louisiana, under the influence of such southern writers as Robert Penn Warren, John Peale Bishop, and Allen Tate, as a literary quarterly. Two previous magazines of the same name were published during the 19th century.

SOUTHWORTH, E[mma] D[orothy] E[liza] N[evitte] (1819-99), wrote more than 60 sentimental novels, while living in Washington, D. C., with frequent reference to the "democratic court life" of the capital. She established one of the most phenomenal "best-selling" records in the history of American publishing over a sixty year period by exploiting the popular feminine taste for artificiality, sensation, declamation, and sentimental melodrama.

SPEYER, Leonora (1872-), New York poet and past President of The Poetry Society of America, was awarded the Pulitzer Prize for *Fiddler's Farewell* (1926).

SPINGARN, J[oel] E[lias] (1875-1939), the most influential American critic in the first two decades of this century, was a disciple of Croce in Impressionistic criticism, which advocated the overthrow of all rules for artists. His definitive book, *The New Criticism* (1911), contained also in *Creative Criticism* (1917), should not be confused with the contemporary "New Criticism" of such writers as Warren, Brooks, and Tate. Spingarn also wrote *Poems* (1924), and *A Spingarn Enchiridion* (1929, a reply to attacks from P. E. More). He was professor of comparative literature at Columbia from 1899 to 1911. (See also, CRITICISM, *Impressionism*.)

SPOFFORD, Harriet [Elizabeth] Prescott (1835-1921), sold innumerable stories, novels, and poems that were extremely popular during her lifetime. Her temporary reputation was first established with the appearance of her story, "In a Cellar" (Feb. 1859), and the poor quality of her work notwithstanding, she was a friend of many prominent New England literary persons, as described in her *A Little Book of Friends* (1916).

SPRAGUE, Charles (1791-1875), received little education, having been apprenticed to a merchant at the age of 13, but by 1829 he had become a respected Boston banker and was chosen as Phi Beta Kappa poet at Harvard. He was influenced by Collins and Gray in much of his work, but he also wrote some poems of domestic New England life that had an original sincerity. His *Writings* were published in a collection in 1841.

SQUIBOB, pseudonym of George Horatio Derby (q.v.).

STAFFORD, Jean (1915-), took her B. A. and M. A. degrees simultaneously at the University of Colorado (1936), and studied for a year at the University of Heidelberg on a fellowship. She taught at Stephens College, did graduate work at Iowa, then went to Boston, the scene of her first novel, *Boston Adventure* (1944), a Proustian study of a poor girl going to work for a rich Boston spinster. *The Mountain Lion* (1947) is about a

boy and a girl escaping from their unattractive home to an uncle's ranch in Colorado, only to face the problem of escaping from each other. *The Catherine Wheel* (1952) is an unusual love triangle of three friends. *Children Are Bored on Sunday* (1953) is a collection of her short stories, most of them printed originally in *The New Yorker*. Miss Stafford has been compared to James for her artistic structure, to the Brontes for her concentrated emotional power, and to Proust for the subtlety of her expression. Whatever her setting, and she has used a wide variety, or whatever the psychological nature of her characters, she works with the sure hand of a master and the intuitive insights of a poet.

STALLINGS, Laurence (1894-), born in Macon, Georgia, was graduated from Wake Forest College in North Carolina. As a journalist in New York, he met Maxwell Anderson and began a collaboration of four plays, the most successful of which was *What Price Glory?* (produced 1924), a realistic drama of World War I which was a sensational success on Broadway and was the first of the many succeeding war plays. *Plumes* (1924), a semi-autobiographical novel of the war, was similar in story to Stallings' popular motion picture, *The Big Parade*. He edited *The First World War* (1933), a photographic account. Recently he has been a news-reel editor and scenarist. (See also, DRAMA).

STANSBURY, Joseph (1742-1809), English-born Loyalist poet, came to Philadelphia when he was 25, and prior to the Revolution, wrote satirical poems concerning the patriots. He took part in the negotiations with Benedict Arnold, and later was forced to flee to Nova Scotia, finally returning in 1793. His poetry was published in a collection of *The Loyal Verses of Joseph Stansbury and Jonathan Odell* in 1860. Unlike the virulent Odell, however, he always maintained a style of wit and pleasant humor in poking fun at the patriots.

STEDMAN, Edmund Clarence (1833-1908), was reared, after his father's death and his mother's remarriage, first by a grandfather and, from the age of 9, by a Puritanical Connecticut lawyer. At Yale he won poetry prizes instead of academic distinction, and was expelled in his sophomore year when he neglected his studies further to tour New England with a friend as a traveling actress. (In 1871 Yale awarded him an honorary degree.) He read law with his uncle, then bought a half interest in the Norwich *Tribune,* and later became a New York journalist. In 1864 he opened a brokerage firm in Wall Street, which he operated successfully while devoting his leisure to the writing of poetry and critical essays, and editing: *A Library of American Literature* (11 vols., 1888-90, edited with Ellen M. Hutchinson); *An American Anthology* (1900); an edition of Poe; and other collections, often in collaboration. In his best known poem, "Pan in Wall Street," he epitomized himself as "the banker poet," and although he was one of the famous poets of his day, his permanent influence stems less from his work than from his stimulation

of interest in American literature, and his generosity to more original poets, including E. A. Robinson. Between 1860 and 1900 he published seven volumes of his own poetry, in addition to his *Poems Now First Collected* (1894).

STEELE, Wilbur Daniel (1886-   ), born in North Carolina, has lived in Colorado, New England, South Carolina, and abroad—all of these regions being reflected in the settings of his novels and short stories, the latter noted for their suspense and vigor and technical excellence. They generally treat the conflicting emotions in human beings, finally plunging to violence and tragedy. There are seven collections: *Land's End* (1918), *The Shame Dance* (1923), *Urkey Island* (1926), *The Man Who Saw Through Heaven* (1927), *Tower of Sand* (1929), *Best Stories* (1946), and *Full Cargo* (1951). For the most part, the nine novels employ the same techniques and subjects: *Storm* (1914), *Isles of the Blest* (1924), *Taboo* (1925), *Meat* (1928), *Undertow* (1930), *Sound of Rowlocks* (1938), *That Girl from Memphis* (1945), *Diamond Wedding* (1950), and *Their Town* (1952). In drama Steele has written *The Terrible Woman and Other One Act Plays* (1925), and, with Norma Mitchell, *Post Road* (1935).

STEENDAM, Jacob (c. 1616-1672), although he wrote in his native Dutch, is sometimes called "the first American poet," having published *Complaint of New Amsterdam in New Netherland* (1659) and *Praises of New Netherland* (1661), during his residence in what is now New York. His work, in translation, was reprinted in the biography by H. C. Murphy (1861).

STEGNER, Wallace [Earle] (1909-), was born in Iowa, but grew up in a number of western states and Saskatchewan. He took his B. A. at Utah, and his M. A. and Ph. D. degrees at Iowa. In 1937 he won the Little, Brown contest for his novelette about Iowa farm life, *Remembering Laughter. One Nation* (1945) won a Houghton Mifflin award, and his stories, collected in *Women on the Wall* (1950), have won two O. Henry Memorial awards. His novels include *On a Darkling Plain* (1940), concerning a war veteran in Saskatchewan; *Fire and Ice* (1941), about a college student deserting Communism under the influence of a wealthy girl; *The Big Rock Candy Mountain* (1943), about a family seeking its fortune on a closing western frontier; *The Preacher and the Slave* (1950), presenting a western labor leader; and *Beyond the Hundredth Meridian* (1954). Since 1945 Mr. Stegner has been professor of English and director of the Writing Center at Stanford University.

STEIN, Gertrude (1874-1946), grew up in the San Francisco Bay area, studied psychology at Radcliffe under William James, brain anatomy at Johns Hopkins, and English literature in London. In 1903 she established her residence in Paris, where her *salon* became famous and over the years attracted Picasso, Matisse, Pound, Ford, Sherwood Anderson, Fitzgerald, and Hemingway. Greatly

interested in experimentation, she used a subjective Naturalism characterized by repetitious and colloquial dialogue in the three stories which she privately printed as *Three Lives* (1909). The poems in *Tender Buttons* (1914) depart from conventional logic and grammar to get at objects themselves. *The Making of Americans* (written 1906-8; published 1925) is a chronicle of her family, presented as an ever recurring generalization of every family, with stylistic devices that maintained a perpetual present in the narrative. *The Autobiography of Alice B. Toklas* (1933) was her own autobiography, presented through the eyes of her secretary and companion whom she took with her to Paris. The book became a Literary Guild selection and an American best-seller. Virgil Thompson composed the music for her opera, *Four Saints in Three Acts* (1934). Her critical essays were published as: *Composition and Explanation* (1926), presenting lectures delivered at Oxford and Cambridge; examples of *How to Write* (1931); *Narration* (1935); and *Lectures in America* (1935). She compared her technique to the frames in a motion picture which present a moving series of instantaneous visions in a rhythmic pattern. The movement derives from the verb. Nouns are names, and can be emphasized only in poetry, which is loving the name of something, but not in narrative prose which needs a forward movement. Through her examples and direct contact with many younger writers, and through their influence in turn on others, she had an enormous influence on contemporary literature, out of all proportion to the quantity or popularity of her own work.

STEINBECK, John [Ernst] (1902-), grew up in the Salinas Valley of California, studied marine biology at Stanford, and then, turning to literature as a career, took a series of laboring jobs to get background. After publishing two novels, *Cup of Gold* (1929), and *To a God Unknown* (1933), and stories collected as *The Pastures of Heaven* (1932), he won both popular and critical acclaim with *Tortilla Flat* (1935), a sympathetic and humorous portrait of Monterey *paisanos* maneuvering to secure wine and bread without degrading themselves by working. This whimsical style was repeated in *Cannery Row* (1944), but meanwhile Steinbeck turned to the social-conscious novel with *In Dubious Battle* (1936), reporting a strike of migratory workers being tragically broken by selfish entrepreneurs. This was an apprentice work for his great, powerful, *Grapes of Wrath* (1939, Pulitzer Prize), a documentary chronicle of a family moving from the Oklahoma dust bowl to the promised land of California, where they find organized animosity, exploitation, and terrible tragedy. The novel suffers from the sentimentality of didacticism, but his powerful symbolism and accurate detail in presenting a national crisis through the experiences of crude people more than compensate for the occasional weaknesses. His other novels include: *Of Mice and Men* (1937), a successful experiment in writing a play with the novel form; *The Moon Is Down* (1942), showing Norwegian resistance to the Nazi occupation; *The Wayward Bus* (1947); *The Pearl* (1948); and *East of Eden* (1952). His story, "The Red Pony," is considered an American classic. Mr. Steinbeck resided in Monterey for many years, and presently lives in New York City.

STEVENS, James [Floyd] (1892-), grew up in Idaho, working for a living from the age of 13. He collected tall tales for his *Paul Bunyan* (1925). His novels include *Brawnyman* (1926), *Mattock* (1927) and *Big Jim Turner* (1948).

STEVENS, Wallace (1879-), was born in Pennsylvania, and after graduation from Harvard, studied law in New York, later joining the staff of the Hartford Accident and Indemnity Company in Connecticut, of which he is now a vice president. His poems were first published in *Poetry* in 1914, and in 1920 he was awarded the Helen Haire Levinson Prize. In 1936 he won the *Nation's* poetry prize with "The Men That Are Falling." His residence in Hartford and his occupation, as well as his aversion to publicity, kept him out of the poetic limelight for most of his creative years, although he has long been recognized by a few critics, including Ezra Pound, as one of the most original and provacative of American poets. Recently his name and work have become more generally known. Using a colorful, impressionistic style reminiscent of the French painters and composers, especially Debussy and Matisse, and a highly intellectual argument that is often difficult to decipher, he writes of the beauty as well as the irony of contemporary life in *Harmonium* (1923, 1931), *Ideas of Order* (1935), *Owl's Clover* (1936), *The Man with the Blue Guitar* (1937), *Parts of a World* (1942), *Notes Toward a Supreme Fiction* (1942), and *Transport to Summer* (1947).

STEWART, George R[ippey] (1895-), professor of English at the University of California, is best known among graduate students of English for his revelations concerning the *Doctor's Oral* (1939), which he took at Columbia (1922) after earning a B. A. at Princeton and an M. A. at California. He has also written the biographical studies: *Bret Harte: Argonaut and Exile* (1931); *John Phoenix, Esquire* (1937); and *Take Your Bible in One Hand* (1931), about William Henry Thomes. *East of the Giants* (1938) is a historical novel about California. His novel using a *Storm* (1941) for the protagonist attracted high praise for his ability to combine meteorology with drama in showing the effect of a low pressure area moving across the country. His other works include an account of the Donner party, *Ordeal by Hunger* (1936); a historical account of *Names on the Land* (1945); *Man, An Autobiography* (1946); a forest *Fire* (1948); two books on prosody, *Modern Metrical Technique* (1922), and *Technique of English Verse* (1930); and a study of the *American Ways of Life* (1954).

STICKNEY, [Joseph] Trumbull (1874-1904), was born in Switzerland, grew up in New England, and after graduation from Harvard, studied the classics for seven years in Paris and Greece, taking the first *doctorat ès lettres* awarded by the University of Paris to an English or American student. His *Dramatic Verses* (1902) and *Poems* (1905) were the finest examples of traditional verse of his day, and had he lived, he most surely would have become one of the great men of American letters. A Harvard friend described him as "the most cultivated man I have ever known." He was one of the first critics to recognize the genius of Edwin Arlington Robinson, and while teaching at Harvard, brilliantly defended Robinson's work.

STOCKTON, Frank R. (Francis Richard) (1834-1902), began his life as an engraver in Philadelphia and New York, then became an author of children's stories, and with the publication of *Rudder Grange* (1879), became a very popular humorist, original in that he depended on fancy rather than dialect for his comic effects. His story, "The Lady or the Tiger" (*Century*, Nov., 1882), created a sensation, and was the title story of a collection issued in 1884. *The Casting Away of Mrs. Lecks and Mrs. Aleshine* (1886), a comic narrative of two widows shipwrecked on a Pacific island, was followed by a sequel, *The Dusantes* (1888). The public demanded sequels also to his first novel, which he provided in *The Rudder Grangers Abroad* (1891) and *Pomona's Travels* (1894). *The Novels and Stories of Frank R.*

Stockton were published in a 23-volume collection (1899-1904).

STODDARD, Charles Warren (1843-1909), was born in Rochester, N. Y., and accompanied his parents to California at 12, returning to study at a New York academy two years later. In 1859 he returned to California, and while clerking in a book store wrote *Poems* (1867), edited by Bret Harte. Later, his travels furnished material for *South Sea Idyls* (1873), *Mashallah!* (1880), *A Cruise under the Crescent* (1898), and *The Lepers of Molokai* (1885). Three years of residence in Hawaii resulted in *Hawaiian Life* (1894) and *The Island of Tranquil Delights* (1904). He described his conversion to Catholicism in *A Troubled Heart* (1885). He served as a professor of English at Notre Dame (1885-6), and at the Catholic University of America (1889-1902).

STODDARD, Richard Henry (1825-1903), his father lost at sea, experienced poverty and squalid living conditions in his youth. He was able to attend a public school between the ages of 10 and 15, and acquired an appetite for literature that consumed his evenings thereafter while he worked as a shop boy, errand boy, and iron molder. After the publication of his *Poems* (1852), Hawthorne helped to secure him a position in the New York customs house (1853-70). Meanwhile he had become a friend of Bayard Taylor and other literary figures of the day. After 1880 he was literary editor of the New York *Mail and Express*, and became an influential

arbiter of conservative taste, conducting with his wife a famous *salon*, visited by Melville as well as the famous writers of the day. He is remembered for his influence and position rather than for his many volumes of poetry, which were imitative in both thought and expression.

STONG, Phil[ip Duffield] (1899-), Iowa-born novelist and scenarist, writes mainly of the small town communities of his native state. *State Fair* (1932) was successful, not only as a novel, but also as a motion picture, with Will Rogers in the leading role. *The Farmer in the Dell* (1935) takes a farmer to Hollywood. In addition to his many novels, he has written humorous books for boys and social studies.

Story Magazine, (1931-), founded by Whit Burnett and his first wife, Martha Foley, in Vienna, and later established in New York in connection with The Story Press. It was the premise of the founders that by using no advertising and by paying each author $25 per contribution regardless of his prominence, they could provide a market for the quality story that would encourage young writers. Their own critical preference was for the story with a social importance, and they were opposed to the surprise ending which O. Henry had brought to such wide popularity. William Saroyan and Tess Slessinger were among their discoveries.

STOWE, Harriet [Elizabeth] Beecher (1811-96), daughter of a Connecticut minister who supported orthodox Calvinism in the Unitarian controversy. In 1832 she moved with her family to Cincinnati where her father ran a Theological Seminary, and in 1836, she married one of the professors, Calvin E. Stowe. Although she visited a Kentucky plantation and was sensitive to the evils of slavery, she did not become an ardent abolitionist until 1850 when, with the passage of the Fugitive Slave Act, her brothers persuaded her to join their cause, and she produced *Uncle Tom's Cabin* (1852). To her amazement, the book became the symbol and the text of the Abolitionist movement, and catapulted her into prominence. On the basis of her accidental fame she became a successful novelist. Her second anti-slavery novel, *Dred, A Tale of the Great Dismal Swamp* (1856) was more accurate and better written than her more famous work. *The Minister's Wooing* (1859) illustrates her more direct attempt to write romantic fiction. Although this work attacks some of the injustices of Calvinism and represents her turning to some degree from the dogmas of her father, she was forever a mixture of the Puritan and the sentimental romantic. She continued to turn out novels, but only the first is remembered.

STRANGE, Robert (1796-1854), North Carolina novelist, wrote *Eoneguski, or the Cherokee Chief* (1839), an alternately romantic and realistic tale of the Cherokee Indians of western North Carolina up to the time of the removal to the Oklahoma territory. In the tradition of Cooper and Simms, it relates the white men's shabby treatment and fre-

quent betrayals of the Cherokees and is based on actual events and real people.

STREET, James (1903-54), born in Mississippi, was a journalist, Baptist preacher, and free-lance writer. After *Look Away* (1936), a book of varied sketches of life in Mississippi, he began a series of novels about the Dabney family in the Mississippi valley: *Oh, Promised Land* (1940), *Tap Roots* (1942), *By Valour and Arms* (1944), *Tomorrow We Reap* (collaboration with James Childers, 1948), and *Mingo Dabney* (1950). *In My Father's House* (1941) is the story of a cotton-farming family in Mississippi; *The Biscuit Eater* (1942), a novelette of two boys and a dog. Street's magazine pieces were collected in *Short Stories* (1945). Two novels concern the unusual career of a Baptist minister in the South: *The Gauntlet* (1945) and *The High Calling* (1950). *The Velvet Doublet* (1953) is a story of Columbus as told by his sailor, Rodrigo de Triana, who first sighted land in the New World. *The Civil War* (1953) is subtitled "An Unvarnished Account of the Late but Still-Lively Hostilities"; its purpose is to shatter favorite myths and romantic notions about the heroes and actions of both sides.

STRIBLING T[homas] S[igismund] (1881-), began his writing career in Tennessee as a successful author of Sunday school stories. After World War I he turned from commercial to more serious fiction, producing a story about race prejudice in *Birthright* (1922). *Teeftallow* (1926), presenting the bigotry in a small Tennessee mountain town, was a best seller and the first selection of the Book-of-the-Month Club. The Literary Guild selected *The Store* (1932; Pulitzer Prize), *The Unfinished Cathedral* (1934), and *The Sound Wagon* (1935). Before writing his novels, Mr. Stribling organizes his story by arranging and rearranging dozens of cards, which he tacks on the wall of a room, until he secures the desired sequence of events.

STUART, Jessie (1905-   ), has written short stories about his native Kentucky, collected in *Head o' W-Hollow* (1936), *Men of the Mountains* (1941), and *Tales from the Plum Grove Hills* (1946). His indigenous Kentucky poems were collected in *Man with a Bull-Tongue Plow* (1934), and *Album of Destiny* (1944), the first book containing 700 sonnets. His first novel, *Trees of Heaven* (1940), contrasts a propertied southern family with a squatter family, their relations being complicated by business and love. *Taps for Private Tussie* (1944) humorously depicts the private's family squandering $10,000 of insurance money after he is reported dead. His other novels are *Mongrel Mettle* (1944), concerning an ambitious dog; *Foretaste of Glory* (1946), sketching a Kentucky town persuaded that the world is about to end; and *Hie to the Hunters* (1950), about a city boy enjoying Kentucky life. *Beyond Dark Hills* (1938) is the autobiography of Stuart, who quit school at eleven to work in the fields, but later worked his way through high school and college.

SUCKOW, Ruth (1892-   ), grew up in many Iowa communities as her father, a Congregational minister, moved to new parishes. She attended Grinnell College, and in 1917 graduated from the University of Denver, where she later taught. Returning to Iowa, she managed an apiary during the summers and wrote during the winters, first attracting attention with her stories that appeared in the *Smart Set* and *American Mercury*. Her first novel, *Country People* (1924), portrays a German American family deteriorating as it gradually grows wealthy during three generations. Family repressions are depicted in *The Odyssey of a Nice Girl* (1925). Her concrete, realistic portraits of Iowa people, who might be people anywhere, include, in addition to those mentioned: *The Bonney Family* (1928) and *New Hope* (1942), both studies of ministers; *Cora* (1929); *The Kramer Girls* (1930); and *The Folks* (1934). Her short stories have been collected in *Iowa Interiors* (1926), and in *Children and Older People* (1931). Miss Suckow and her author husband, Ferner Nuhn, now live in Arizona.

SUT LOVINGOOD; see G. W. Harris.

T

TABB, John B[anister] (1845-1909), a Confederate blockade runner during the Civil War, was captured and imprisoned at Point Lookout (1864), where he met Sidney Lanier, who encouraged him in the writing of poetry. Later Tabb was converted to Catholicism and ordained as a priest, and from 1884 until his death taught English at St. Charles's College near Baltimore. His *Poems* (1894), *Lyrics* (1897), *Later Lyrics* (1902), and *The Rosary in Rhyme* (1904) have been compared to Donne in their religious intensity, and to Emily Dickinson for their frequently epigrammatic style.

TAGGARD, Genevieve (1894-), grew up in Hawaii, graduated from the University of California (1919), and after serving in editorial positions in New York, taught English at Mount Holyoke, Bennington College, and Sarah Lawrence College. *For Eager Lovers* (1922), her first volume of poems, contained fairly conventional, personal lyrics, but in *Words for the Chisel* (1926) and *Travelling Standing Still* (1928), she reflected the new vogue for metaphysical poetry. *Calling Western Union* (1936) revealed her interest in social issues. Her *Collected Poems* (1938) were followed by *The Long View* (1942) and *Slow Music* (1946). She has also written a prose study of *The Life and Mind of Emily Dickinson* (1930).

TARBELL, Ida M[inerva] (1857-1944), was a leader of the muckraking school of journalism at the turn of the century. Her articles written for *McClure's* magazine were edited for publication of *The History of the Standard Oil Company* (2 vols., 1904), one of the first of the sensational exposés of private enterprise. In addition to works on Lincoln and other works on business, she wrote her autobiography, *All in the Day's Work* (1939).

TARKINGTON, [Newton] Booth (1869-1946), an Indiana neighbor and friend of James Whitcomb Riley, was educated at Purdue and Princeton, and after abandoning his ambition to become an illustrator, won prominence for *Monsieur Beaucaire* (1901), an historical novel of the 18th-century Duke of Orleans. Although his first novel, *The Gentleman from Indiana* (1899), had been a commercial failure, he now felt secure in again portraying the life of the Middle West in a series of novels, climaxed by *The Magnificent Ambersons*

(1918), and *Alice Adams* (1921), both taking the Pulitzer Prize. His novel of the boy, *Penrod* (1914), is one of his best known works. A bibliography of his many novels, plays, short stories, and essays runs to several pages.

TATE, [John Orley] Allen (1899-   ), one of the most influential of the "New Critics," graduated from Vanderbilt University in 1922, and became a founder and editor of *The Fugitive,* a literary journal. He has been a critic and reviewer for several magazines and has lectured at numerous colleges. His poems, marked by a metaphysical tension of polished style and satirical expression, have been collected in *Mr. Pope and Other Poems* (1928), *Three Poems* (1930), *Poems, 1928-1931* (1932), *The Mediterranean and Other Poems* (1936), *Selected Poems* (1937), and *Winter Sea* (1945). His criticism includes *Reactionary Essays on Poetry and Ideas* (1936), *Reason in Madness* (1941), and *On the Limits of Poetry* (1948). He has also edited anthologies, written biographies of Stonewall Jackson and Jefferson Davies, and published a novel, *The Fathers* (1938). (See also CRITICISM, The New Critics).

TAYLOR, Bayard (1825-78), poet, editor, novelist, and traveling correspondent, rebelled against his Pennsylvania Quaker family, and at the suggestion of R. W. Griswold, contracted to write travel essays for the *Saturday Evening Post,* the *U. S. Gazette,* and the New York *Tribune.* The spontaneous sketches which he sent to these journals during a two-year tour of Europe (1844-6) were published as *Views A-foot* (1846). His next assignment, to cover the California gold rush, resulted in publication of *Eldorado* (2 vols., 1850). Meanwhile his conventionally romantic poems were published as *Ximena* (1844), *Rhymes of Travel, Ballads and Poems* (1849), *A Book of Romances, Lyrics, and Songs* (1852), and *Poems of the Orient* (1855). With brief intervals of hack writing and editing, he continued as a travel writer, the variety of his journeys being reflected in such titles as *A Journey to Central Africa* (1854); *The Lands of the Saracen* (1855); *A Visit to India, China, and Japan, in the Year, 1853* (1855); and *Travels in Greece and Russia* (1859). In 1862 he served as secretary of the legation in St. Petersburg, and he spent the following seven years mainly in translating Goethe's *Faust.* He died in Berlin a few months after his appointment as minister to Germany. Tremendously popular and highly respected in his own time, Taylor is remembered now as the laureate of the gilded age. A writer whose creative energy far exceeded his ability, his very success in an age of gilded values indicates the hollowness of his work.

TAYLOR, Edward (c. 1644-1729), emigrated from England to Boston at the age of 24, and after study at Harvard, became a pastor at Westfield, Massachusetts. His religious poems were passed down in manuscript through several generations of his descendants until 1939, when they were published in a volume derived from his papers in the Yale library. He was immediately recognized to be a literary kin of the 17th-century English metaphysical poets, and by far the most important of the early American poets. His devotional poems have an honest simplicity of statement set in precise verse.

TEASDALE, Sara (1884-1933), grew up in St. Louis, where she attended a private school, and spent two years (1905-7) in Europe and the Near East. After her marriage in 1914 to an exporter, she lived in New York City. Her poems began to appear in the leading magazines during her twenties, and were collected as *Sonnets to Duse and Other Poems* (1907); *Helen of Troy* (1911); *Rivers to the Sea* (1915); *Love Songs* (1917), which received a special Pulitzer award; *Flame and Shadow* (1920); *Dark of the Moon* (1926); *Strange Victory* (1933); and *Collected Poems* (1937). Her brief, feminine lyrics of deep personal experience were distinguished by a sparseness of image and metaphor, and by a richness of sensitivity and dexterity.

TENNEY, Tabitha [Gilman] (1762-1837), the wife of a U. S. Congressman from New Hampshire, wrote a satire on the feminine taste of her day for romantic novels, *Female Quixotism* (2 vols., 1801). A bookseller's advertisement announced her as the editor of an anthology of poetry and classical essays, *The Pleasing Instructor,* but no copy of this work has been found.

TERHUNE, Albert Payson (1872-1942), a popular novelist whose major protagonists were super-canine collie dogs: *Lad, A Dog* (1919), *Bruce* (1920), and *Lad of Sunnybank* (1928), to mention a few. The books formed a saga of several generations of collie champions.

TERHUNE, Mary Virginia (1830-1922), under the pseudonym of Marion Harland, wrote 26 popular romantic novels, but achieved an even greater success and made a more lasting contribution with *Common Sense in the Household* (1871), one of the first books on home economics. This work caused a great demand for additional advice concerning home management, and she produced many articles and books on the subject. Her son, Albert Payson Terhune, was also a successful novelist.

Theatre Arts Monthly, founded in 1916 as *Theatre Arts Magazine,* the change in the name taking place in 1924 when it changed from a quarterly to a monthly publication. The magazine publishes both news and criticism concerning the American theater, and has maintained a progressive attitude toward experimental productions.

Theatre Guild, The, was founded as a little-theater by members of the Washington Square Players, but achieved such great financial success that it was able to build its own theater in 1925, and has since become one of the leading producers of successful plays. The favorable reception of Shaw and O'Neill by New York audiences has contributed considerably to the Guild's success.

THOMPSON, Daniel Pierce (1795-1868), Vermont novelist and lawyer, who is remembered principally for *The Green Mountain Boys* (1839), an early regional novel dealing with the New Hampshire land grants, and with Ethan Allen. Pierce's many historical and regional novels were written in the Scott tradition of romance with Cooper settings.

THOMPSON, [James] Maurice (1844-1901), the son of an Indiana Baptist minister, served in the Confederate army, and later became an engineer and a lawyer. His dialect poems were published as *Hoosier Mosaics* (1875), and his most popular novel was *Alice of Old Vincennes* (1900), which related the history of the Northwest Territory and the explorations of George Rogers Clark. Many of his works appeared in the *Atlantic*, and in his day he was considered a leading literary figure of the Middle West.

THOMPSON, William Tappan (1812-82), a Georgia journalist, wrote dialect stories humorously portraying the Georgia cracker, collected in *Major Jones's Courtship* (1843).

THOMSON, Mortimer Neal (1831-75), who wrote under the pseudonym of Q. K. Philander Doesticks, P.B., was a New York journalist and humorist. His satirical newspaper sketches were collected as *Doesticks, What He Says* (1855). He also wrote parodies, including *Plu-ri-bus-tah, A Song That's By No Author* (1856), aimed at Longfellow's *Hiawatha*.

THOREAU, Henry David (1817-62), the Concord essayist, was born of mixed French and Scottish parentage, his father having come from the Isle of Jersey. As a boy, Thoreau helped his father in the manufacture of pencils, and spent his spare time in nature study. When he graduated from Harvard (1837) he refused to spend five dollars for the diploma. After teaching in a school run by his brother (1837-1841), he became a disciple and handyman of Emerson, and joined the Transcendentalist group. He spent the year of 1843 on Staten Island, New York as a tutor at the home of William Emerson, a brother of Ralph. In 1845 he built himself a house on Walden Pond for $28, and lived there in studied communication with nature, surveying or carpentering for his small expenses. There he wrote *A Week on the Concord and Merrimac Rivers* (1849), privately publishing a thousand copies, of which only 200 were sold. His next book, *Walden* (1854), had a greater success, and he received invitations to contribute to the journals, and to deliver lectures, but he was hardly the man to capitalize on success, being a quite rugged individualist whose opinions were never popular in his own time. A strong Abolitionist, he was jailed for refusing to pay poll tax to a government that allowed slavery (1845), and he was the first public defender of John Brown. In 1847 he left his hermitage at Walden Pond, served as a caretaker for Emerson, and took long walking tours even as far as Canada. When his father died in 1859, he and a sister ran the pencil factory together. In 1861 a cold developed into tuberculosis, and when a trip to

Minnesota failed to brace him as he had hoped, he came home to Concord to die. His letters and extracts from his journals were published posthumously in a series of books during the next fifty years, and although practically unknown in his lifetime, he has come to be regarded as a major New England writer. His verse, imitative of Emerson, was mere scholarship in doggerel, but his prose revealed an unusual wit and a vigorous independence of conventional attitudes. A mixture of the practical Yankee and the mystic Transcendentalist, and also something of an anarchist as he expressed his views in "Civil Disobedience" (1849), his work is characterized by flashes of poetry or original observation written in a nervous, staccato style. He was a man who had "travelled a good deal in Concord," seeking wisdom in his own mind.

THORPE, T[homas] B[angs] (1815-78), a Louisiana journalist and humorist, first exploited the "tall tale" in regional humor. His best work is "The Big Bear of Arkansas," which was first published in *The Spirit of the Times* journal, and was later included in *The Mysteries of the Backwoods* (1846).

THURBER, James [Grover] (1894-), grew up in Columbus and graduated from Ohio State University, and after serving as a code clerk in the Paris embassy (1918-20), became a newspaper reporter in Columbus, Paris and New York. In 1926 he joined the staff of *The New Yorker,* in which his work has regularly appeared. He is as beloved for his drawings of limp people and soulful dogs as for his humanly humorous stories

and essays. With E. B. White, whom he credits with teaching him the art of writing, he wrote *Is Sex Necessary?* (1929), satirizing pseudo-scientific sex articles. His other work includes: *The Owl in the Attic and Other Perplexities* (1931); *The Seal in the Bedroom and Other Predicaments* (1932); *My Life and Hard Times* (1933); *Fables for Our Time* (1940); *Men, Women, and Dogs* (1943); and the collection of his selected work, *The Thurber Carnival* (1945). With Elliot Nugent he wrote the highly successful comedy, *The Male Animal* (1940), which satirizes collegiate emphasis on athletics and other non-intellectual mores. His story, "The Secret Life of Walter Mitty," was made into a motion picture. This and many other stories, such as "The Dog that Bit People," have become American story classics, and are frequently included in anthologies. Thurber achieves a deft and subtle humor through whimsy and personal characterization rather than from word manipulation or cleverness, and much of his rare quality stems from revelation.

TIERNAN, Mrs. Frances Christine Fisher (1846-1920), a prolific romantic author of the postwar South, wrote more than forty novels under the pseudonym of "Christian Reid." Poems and short stories also came easily from her pen. Though her characterizations were stilted, she was gifted as a descriptive writer. *Valerie Aylmer* (1870) and *The Land of the Sky* (1876) are among her best-remembered novels.

TIMROD, Henry (1828-67), grew up in

near-poverty after his father's death in 1838. He spent two years at Franklin College, tried the law profession, and became a tutor. Contributing poems to the *Southern Literary Messenger* and *Russell's Magazine,* he became a member of the Charleston literary group which included Simms and Hayne, and published a volume of *Poems* in 1860. Suffering from tuberculosis, he was too ill to serve with the Confederate Army during the Civil War, and after a year of clerking duty, was discharged. He relieved his intense fervor for the southern cause by writing patriotic poems, and came to be known as "the laureate of the Confederacy," but poetry could not abate the poverty and suffering which the war brought to his life, and he died before his poetic talents could be fully realized. Hayne published his collected *Poems* in 1873; followed by *Katie* (1884), a long love lyric Timrod had written for his English wife; and *Complete Poems* (1899). Although some of his poems suffered from patriotic didacticism, he achieved lyric strength in his "Ode" on the Confederate dead, and in "The Cotton Boll."

TOCQUEVILLE, Alexis, Comte de (1805-59), a liberal French aristocrat whose analysis of American society and culture was so objectively penetrating that he is generally included in studies of American literature even though he spent only nine months in the U. S. (1831-2). Sent by the French government to study our penitentiary system, he wrote a two volume study of democracy upon his return to France, *De la Démocratie en Amérique* (1835; American edition, 1838; 2 supplementary volumes, 1840). Although he praised many aspects of American democracy, he saw dangers in collective despotism, and feared that our great emphasis on equality might endanger personal liberty. In literature this equalitarianism would substitute universal man for heroic man, and both profundity and subtlety of expression would be replaced by originality and vigor.

TODD, Mabel Loomis (1856-1932), author, wife of an Amherst professor, and good friend and neighbor of Emily Dickinson, collaborated with T. W. Higginson in the editing of the first and second posthumous collections of Miss Dickinson's *Poems* (1890-91). She also edited a third edition (1896), and the *Letters of Emily Dickinson* (2 vols., 1894). Later a property dispute with the Dickinson family terminated her editorial endeavors.

TOKLAS, Alice B.; see Gertrude Stein.

TOMPSON, Benjamin (1642 - 1714), teacher and physician, is remembered as the first native-born Colonial poet, and not particularly for the quality of his satire on King Philip's War, *New England Crisis* (1676).

TORRENCE, [Frederick] Ridgely (1875-1952), New York poet and dramatist, and an early friend of Robinson, expressed the popular sentimentality of the time in *The House of a Hundred Lights* (1900), but showed a philosophic maturity in *Hesperides* (1925), and *Poems*

(1903) and *Abelard and Heloise* (1907). (1941). His plays include *El Dorado*

TOTHEROH, Dan (1895-   ), a California dramatist and director, whose *Wild Birds* won a University of California contest and was later produced in New York (1925).

TOURGÉE, Albion Winegar (1838-1905), native of Ohio and student at the University of Rochester, served as a lieutenant in the Union Army, was twice wounded, then in 1865 went to North Carolina and entered politics during the carpetbagger regime. A man of considerable ability and intelligence, he was nevertheless violently partisan and as a Superior Court Judge was not guiltless of participation in corrupt administration. He moved to New York State in 1879. Though a prolific writer, he is remembered for a series of novels reflecting the Republican attitude toward the postwar South. They were vastly popular in their day and are still important for their biased social and political views. *Toinette* (1874), revised as *A Royal Gentleman* (1881), is the tragic story of a southern planter who loved but would not marry his slave girl. *Figs and Thistles* (1879) follows the partly autobiographical career of an Ohio politician during the Civil War. *A Fool's Errand* (1879), his most admired work, is based on the author's life in North Carolina during Reconstruction, as is *Bricks Without Straw* (1880). *John Eax and Mamelon* (1882), novelettes published together, deal with the postwar South. *Hot Plowshares* (1883) is a race novel

set in New York State, while *Pactolus Prime* (1890) tells of a Negro's "passing" the race barriers in Washington.

Transcendentalism, a name derived from Kant's *Critique of Practical Reason* to define a humanistic, idealistic mysticism which was the practical philosophy of many New England writers in the mid-19th century under the leadership of Emerson. Kant defines transcendental knowledge as a mode of knowing objects, and his philosophy influenced other German philosophers, such as Herder, Jacobi, and Schelling, who in turn influenced the English Coleridge, Wordsworth, and Carlyle, who influenced the New Englanders. But the New England brand was a mixture of Jonathan Edwards, Goethe, Plato, Confucius, Sanskrit religion, Buddhism, and Swedenborg. It emerged as a monistic philosophy which held that man or nature or any object in the world is a microcosm containing the macrocosm of God. Practically, the cult emphasized individualism and self-reliance as opposed to tradition and authority, and feeling as opposed to logic. The individual's Over-Soul, identical with the soul of God, was his source of morality, divinity, and beauty. The Transcendentalists were not formally organized, and the philosophy varied in the presentations of its followers. Besides Emerson, Thoreau presented the clearest expression of the attitude in his *Walden*. Other Transcendentalists were Alcott, the Channings, Margaret Fuller, Brownson, and Elizabeth Peabody. Hawthorne, Longfellow, Bryant, Whittier, and even Whitman and Melville, although not active in the cult, were affected by its doctrines.

TRAUBEL, Horace L[ogo] (1858-1919), a Camden editor and author who became an intimate friend of Walt Whitman and wrote *With Walt Whitman in Camden* (3 vols., 1906-14), a journal of his visits with the poet after 1888, with some criticism and conjecture as well as fact. He also edited *In Re Walt Whitman* (1893), and *The Complete Writings of Walt Whitman* (10 vols., 1902). As a source for later scholars his work on Whitman had considerable value, but he is now studied mainly by specialists.

TRILLING, Lionel (1905-), grew up in New York City, and took his A.B. (1925), M.A. (1926), and Ph.D. (1936) degrees at Columbia, where he is now a Professor in the Graduate English department. He taught at the University of Wisconsin in 1926-27, and at Hunter College from 1927 to 1930. He has won distinction as a scholar, a lecturer and a novelist. He wrote the definitive study of Matthew Arnold (1939), and a critical examination of E. M. Forster (1943). His critical essays were collected in *The Liberal Imagination* (1950). *The Middle of the Journey* (1948) is a novel. He also has edited *The Portable Matthew Arnold* (1949) and *The Letters of John Keats* (1950). His criticism is distinguished by fresh insights into the work of each subject he studies, and although he has worked closely with the New Critics (q.v.), he is an independent scholar.

TRUMBULL, John (1750-1831), one of the Connecticut Wits, learned Greek and Latin at five and passed the entrance examinations at Yale College at seven, although he was not permitted to enter until he was thirteen He took his B.A. at seventeen and his M.A. at twenty. While tutoring at the college he wrote *The Progress of Dulness* (1772-3), a satirical attack on the curriculum. In 1773 he moved to Boston to study law in the office of John Adams, and was drawn into the patriotic movement, which led him to write the bombastic *An Elegy on the Times* (1774), and the mock epic burlesque of Tory politics, *M'Fingal* (1782). After the Revolutionary War he lived in Hartford and became Federalist in his aristocratic sympathies. With other conservative poets he wrote *The Anarchiad,* satirizing Jeffersonian democracy. In later life his poetic powers completely declined, and he became a prominent local jurist.

TUCKER, George (1775-1861), Bermuda-born Virginia legislator and professor at the Uuniversity of Virginia (1825-45), wrote two novels, *The Valley of Shenandoah* (1824) and *A Voyage to the Moon* (1827), as well as many economic studies including *Progress of the United States in Population and Wealth in Fifty Years* (1843).

TUCKER, Nathaniel Beverley (1784-1851), the son of St. George Tucker, grew up in Virginia and graduated from William and Mary, where he later became a professor of law. In addition to many books and essays written in defense of the reactionary Southerner's political attitudes, he wrote three novels: *George Balcombe* (1836), a fairly realistic delineation of life in Virginia and Missouri (where he served as a circuit court judge), praised by Poe as the best American novel ever written; *The Partisan Leader* (1836) a frequently accurate prophesy of the future political history, post-dated 1856 and written as a historical romance; and *Gertrude* (1844-5), a novel in which his verbosity and sentimentality are unrelieved by such virtues as are found in his earlier stories.

TUCKER, St. George (1752-1827), was born in Bermuda, and after attending William and Mary College, settled in Williamsburg, Virginia to practise law. He served in the Revolutionary War with distinction, achieving the rank of lieutenant colonel. Although his most important writing was in the field of politics and law, including a *Dissertation on Slavery* (1796), suggesting gradual emancipation, his poems also had considerable merit, and included *Liberty, a Poem on the Independence of America* (1788); and the political satires, *The Probationary Odes of Jonathan Pindar* (1796).

TUCKERMAN, Frederick Goddard (1821-73), although trained at Harvard for a legal career, was able to retire in 1847 to Greenfield, Mass., to devote himself to private botanical studies and to the celebration of nature in his poetry. A few of his poems were published in *Living Age* and the *Atlantic,* and a volume of his *Poems* in 1860 warranted three editions. Emerson and Longfellow praised his work, as did Tennyson, with

whom he stayed during a visit to England. Yet, he did not write specifically in the romantic tradition; he took unpopular liberties with the sonnet form, and attained very little success in his lifetime. After his death he was completely forgotten until Walter Pritchard Eaton called attention to his work in 1909, and in 1931 Witter Bynner published a collection of his best works, praising them for their radical prosody and noble expressions of grief.

TUCKERMAN, Henry Theodore (1813-71) a sentimental poet and sympathetic critic whose cultural effervescence caused his contemporaries to over-estimate his literary ability. His works included *The Italian Sketch Book* (1835), *The Optimist* (1850), and *The Criterion* (1866).

TURNBULL, Belle was born in Hamilton, New York, and graduated from Vassar. She has lived most of her adult life in Colorado, as head of the English Department of the Colorado Springs High School, and as a free-lance writer living in the high mountain town of Breckenridge. Her poems have been published in *Poetry* and other magazines. *Goldboat* is a verse novel, and *The Far Side of the Hill* (1953) is a prose novel about Colorado mountain country that is told with poetic overtones of meaning.

TWAIN, Mark (1835-1910), the pen name of Samuel Langhorne Clemens, whose birth in Florida, Missouri was heralded by the appearance of Halley's Comet. John Marshall Clemens had come west seeking the promise of *The Gilded Age* in land speculation. In Kentucky he had married Jane Lampton, and in Tennessee he had bought eleven thousand acres of virgin land for a few hundred dollars. The town of Florida, he was sure, would become a western metropolis when the Salt River had been dredged. After waiting four years for Congress to pass the dredging bill, the Clemens family moved east to Hannibal, Missouri, destined to become famous as the setting for their youngest son's novels of his boyhood. By the time John Clemens died in 1847, the family had found only poverty in the West, but he urged them on his deathbed to "Cling to the Tennessee land, it will make you all rich." (Ironically, after the family sold it many years later, the land was discovered to be underlaid with coal.) At 14 Samuel was apprenticed to the publisher of the *Missouri Courier* as a printer. Two years later he transferred to the Hannibal *Western Union,* published by his brother, Orion, and in addition to printing he sometimes wrote for the paper. At 18 he became a tramp printer, working in New York, Philadelphia, Washington, Cincinnati, St. Louis, and various smaller towns. In 1857 he left Cincinnati for New Orleans on a river boat, planning a trip to South America, but the river journey recalled his boyhood dream of becoming a river pilot. He managed to persuade the pilot, Horace Bixby, to teach him the Mississippi river for $500, and the arduous training, later described in *Life on the Mississippi* (1883), took a year and a half. He served then as a licensed pilot on the romantic, adventurous river, an unquestioned hero to small boys and half the men, until the Civil War closed

the Mississippi to navigation. After a burlesque service with a group of Confederate volunteers, he went to Nevada with his brother, who had been appointed secretary to the governor. He tried unsuccessfully to make a fortune prospecting, then became a reporter in Virginia City (1862), where he adopted the pseudonym, Mark Twain, a river phrase meaning "two fathoms deep." It had previously been used by another pilot, Isaiah Sellers, in writing for a New Orleans paper pompous articles that Clemens had burlesqued. Twain's western trip and later adventures are reported in *Roughing It* (1872). In 1864 he began working for papers in San Francisco, sometimes collaborating with Bret Harte, and in 1865 he first won national recognition with his story, "The Celebrated Jumping Frog of Calaveras County." After returning from a reporting trip to the Sandwich Islands, he lectured in California on his adventures and observations, and discovered that he had the rare ability to entertain an audience. His next adventure was a tourist journey on the *Quaker City* to the Mediterranean and the Holy Land, which he described in letters sent back to California papers. The passengers included many persons of culture and even scholarship with whom Twain became fairly intimate, and they, in combination with a first hand view of European centers of culture, affected him profoundly. Although he was captivated by the financial opportunities that developed upon his return, he edited his newspaper reports thoroughly before publishing them as *Innocents Abroad* (1869). This was still the incomparable Twain, "The Humorist of the Pacific Slope" pricking the bubble of cultural ostentation with the pin of practical wit, but he was now a refined Twain that would appeal to English professors as well as to hard rock miners. The book was a great success, and in 1870 he married Olivia Langdon, of a genteel and conservative family. By 1872 they had taken a home in Hartford, and Twain settled down to serious writing under the influence of William Dean Howells, Thomas Bailey Aldrich, and Mrs. Clemens. A critical controversy has developed as to whether their influence was repressively harmful, as maintained by Van Wyck Brooks, or repressively beneficent, as Bernard DeVoto insists. A comparison of his California work with that done at Hartford has led most scholars to support the DeVoto position. With a Hartford neighbor, C. D. Warner, he wrote *The Gilded Age* (1873). In *A Tramp Abroad* (1880), he reports a walking trip through the Alps with less spontaneous charm than he displayed in *Innocents Abroad*. *The Prince and the Pauper* (1882), set in England during the reign of Edward VI; *A Connecticut Yankee in King Arthur's Court* (1889), a fantasy contrasting American ingenuity, aided by modern inventions, with Merlin's magic; and *Joan of Arc* (1896); are all attempts to produce "literature." Insofar as Howells or Mrs. Clemens were responsible for these attempts, Mr. Brooks has a point. But Twain also wrote his greatest work after he had come under their influence. *Tom Sawyer* (1876), his classic portrait of American boyhood, is hung on a routine, adventure

plot, but the insights into adolescent psychology which emerge as sympathetic and rare humor have never been equalled. *The Adventures of Huckleberry Finn* (1884), now regarded as his greatest work, and by some critics as the greatest American novel, gave him difficulty, and he set it aside in mid-stream of the Mississippi for several years. Here adolescent psychology and humor are interlaced with overtones of tragedy, satire, and profound social criticism, and strengthened by the symbol of the river. Never has the dichotomy between Chritian ideals and slavery so forcefully been satirized, or the subtleties of childhood maturity been so sensitively contrasted with adult immaturity. In his later life Twain suffered financial reverses from unwise publishing ventures and the failure of a typesetting machine in which he had invested $200,000. Bitter from the death of a daughter (1896) and Mrs. Clemens (1904), Twain produced one more powerful work, *The Mysterious Stranger* (1916), a cynical blast at human values with Satan for his protagonist. He used the same satirical machinery for this novel that earlier had produced laughter, but the reader does not laugh. Many honors were bestowed upon him, including honorary degrees from Yale and Oxford. In 1910, with the reappearance of Halley's Comet, he died. Despite the great volume of critical attention he has received, he remains an enigma in American letters, for the critics have never been reconciled that a man who could at times write so badly could at other times—or even at the same times—reveal such a wealth of genius.

He professed to be interested only in money, and had he restricted his fortune-seeking to writing he would have died wealthy. Had he kept to his original money-making style and avoided what he conceived to be literature, the mean quality of his work would have been higher. But whatever forces shaped Mark Twain, it is doubtful that any critic playing God, and given the power to rearrange those forces in 1835, could have created a greater writer.

TWICHELL, Joseph Hopkins (1838-1918), a Hartford Congregational minister and friend of Mark Twain. He is the prototype of "Harris" in *A Tramp Abroad,* and it was he who suggested that Twain write *Life on the Mississippi.*

TYLER, Royall (1757-1826), graduated from Harvard in 1776, served in the Revolutionary War, and worked in the law office of John Adams, who sent his daughter, Abigail, to France in order to break her engagement with the young man. Tyler first won literary fame with his play, *The Contrast* (1787), the first American comedy and the first to introduce Yankee dialect as a comic device. He later settled in Vermont where he became the Chief Justice of the state's Supreme Court. In addition to other plays he wrote satirical verse; the picaresque novel, *The Algerine Captive* (1797); and a series of fictional letters of a *Yankey in London* (1809).

U

UNTERMEYER, Jean Starr (1886-   ), former wife of Louis Untermeyer, wrote poems that were collected in *Love and Need* (1940).

UNTERMEYER, Louis (1885-   ), left high school in New York at 17 to enter his father's jewelry-manufacturing business, and retired twenty years later as Vice President to devote himself to literature. His romantic poems have been published in more than ten volumes including *Roast Leviathan* (1923), *Burning Bush* (1928) and *Food and Drink* (1932). His *Collected Parodies* (1926) include witty burlesques of modern popular poets and paraphrases of Horace as other poets would have written his works. He wrote the novel, *Moses* (1928); a biography of *Heinrich Heine* (1937); and an autobiography, *From Another World* (1939). He is perhaps best known to the general reader for his poetry anthologies, numbering more than a dozen, of British and American poems.

V

VAN DINE, S. S., see W. H. Wright.

VAN DOREN, Carl [Clinton] (1885-1950), the son of a country doctor in Illinois, graduated from the State University, and took his Ph.D. at Columbia, where he later became an instructor. He served as literary editor of the *Nation* and the *Century*, and in 1926 he was one of the founders of the Literary Guild. His biographical study of *Benjamin Franklin* (1938) was awarded the Pulitzer Prize. In addition to numerous scholarly accomplishments, he wrote a novel, *The Ninth Wave* (1926); stories, collected in *Other Provinces* (1925); and an autobiography of his *Three Worlds* (1936). Irita Van Doren was his wife, and Mark Van Doren his brother.

VAN DOREN, Mark [Albert] (1894-), took his B.A. and M.A. degrees at the University of Illinois and his Ph.D. at Columbia, where he is now a professor of literature. His *Collected Poems* (1939) were awarded the Pulitzer Prize, and his critical studies include Thoreau, Dryden, and Shakespeare. He has also written books for children, and a novel, *The Transients* (1935).

VAN VECHTEN, Carl (1880-), journalist, novelist, and music critic, grew up in Iowa, graduated from the University of Chicago (1903), and became a music critic for *The New York Times*. In addition to musical studies and books about cats, he has written several novels, including *Peter Whiffle* (1922), *The Blind Bow-Boy* (1923), *Spider Boy* (1928), and *Parties* (1930), all describing with a sophisticated wit the elegant lost generation. *The Tattooed Countess* (1924) is set in Iowa, and *Nigger Heaven* (1926) is a realistic study of Harlem. *Sacred and Profane Memories* (1932) is a collection of autobiographical essays.

Veritism, a term used by Hamlin Garland in his *Crumbling Idols* (1894) to define a new realism which went beyond the theories of William Dean Howells to include an examination of the ethical and metaphysical nature of reality. He advocated the exploration of truth to its underlying meaning, a revelation of the acutely felt and immediately expressed moment of experience. He advocated a Realism motivated by democratic purposes, which would explore the unpleasant as well as the pleasant aspects of life. Veritism was thus a pioneer for Naturalism (q.v.) in America, and Garland's theory gave an early support to such Realists as E. W. Howe, Harold Frederic, Stephen Crane, and Henry Blake Fuller. As the new Realism moved more definitely into the channels of Naturalism, Veritism was abandoned as a literary definition.

VERRAZZANO, Giovanni da (c. 1480-1527?), Florentine corsair and navigator, visited the North American coast in 1524 in the employ of Francis I of France. He explored from North Carolina to Maine, and is said to have been the first white man to enter New York harbor. His letter to Francis, a vivid and occasionally beautiful report on the land, includes some exciting encounters with the natives, and is the first description of the continent north of Florida.

VERY, Jones (1813-80), a New England Transcendentalist, poet, and critic, was the son of a sea captain who died when Very was 14. Although forced to earn his living thereafter, he managed to receive a Harvard education. His *Essays and Poems* (1839) were edited by Emerson. His work was more mystical than that of the other Transcendentalists, and has been compared rather to the metaphysical poetry of the 17th-century.

VESTAL, Stanley (1887-) the pseudonym used by Walter Stanley Campbell in writing poetry, fiction, biography, and history of the southwestern frontier, which include: *Ballads of the Old West* (1927); *Kit Carson* (1928), *Happy Hunting Grounds* (1928), *'Dobe Walls* (1929), *Sitting Bull* (1932), *Warpath* (1934), *King of the Fur Traders* (1940), *Short Grass Country* (1941), and *Bigfoot Wallace* (1942). He also wrote *The Missouri* (1944) for the "Rivers of America" series.

VIERECK, Peter (1916-), graduated from Harvard in 1937, where he also earned his M.A. and Ph.D. degrees, the latter in 1942. His speciality is history, and he now teaches at Mount Holyoke. His collection of poems, *Terror and Decorum*, was awarded the Pulitzer Prize in 1949, and was followed by *Strike Through the Mask*. He has also written two historical works, *Metapolitics,* and *Conservatism Revisited.*

Vorticism, a movement developed in London by Ezra Pound with the assistance of Wyndham Lewis after Imagism, which he had developed earlier, had been monopolized by Amy Lowell. With Lewis, Pound published *Blast* (1914-15), the Vorticist manifesto. Closely akin to Imagism, Vorticism refined his definition of the desired image as something that would resemble, in concentration, the spiral of water which develops in the drain of a wash bowl. Lacking an Amy Lowell to organize and publicize its tenets, the new movement quickly perished of neglect.

W

WALLACE, Lew[is] (1827-1905), son of the governor of Indiana, after learning law in his father's office, served in the Mexican and Civil Wars, and reached the rank of major-general. He was the appointed governor of New Mexico (1878-81) and minister to Turkey (1881-85). Although his father had taken him out of school at 16 because of a lack of scholarly application, he had been interested in literature since boyhood, and the success of his first book concerning the Spanish conquest of Mexico, *The Fair God* (1873), encouraged him to try a historical novel concerning the rise of Christianity in the Roman Empire. *Ben-Hur* (1880) became one of the best-selling novels in American publishing history, and long after Wallace's death was a popular movie. His other works, *The Boyhood of Christ* (1888), and *The Prince of India* (1893), are now barely remembered.

WALLACE, William Ross (1819-81), a Kentucky-born New York lawyer and minor poet who was popular in his day, was an intimate friend of Poe and defended him against John Neal's attacks. His poem, "The Hand that Rocks the Cradle," is still published in some anthologies.

WALTER, Eugene (1874-1941), the author of popular melodramas, dramatized *The Trail of the Lonesome Pine* (1912) and *The Little Shepherd of Kingdom Come* (1916).

WARD, Artemus (1834-67), the pseudonym by which Charles Farrar Browne is better known, was apprenticed to a printer at 12 when his father died, and like Twain, two years later shifted to his older brother's paper, the Norway (Maine) *Advertiser*. He became famous as Artemus Ward while working on the Cleveland *Plain Dealer* when, in 1857, he began writing a series of letters purporting to come from the manager of a side show, liberally sprinkled with the puns, misspellings, and poor grammar that delighted readers of his day. After serving with *Vanity Fair*, he became a popular lecturer, and while touring California and Nevada, spent a gay week with Mark Twain, and sent Twain's jumping frog story to a New York editor, thus launching the career of his fellow humorist. On his London tour he delighted even the Queen, but he died there of tuberculosis at the height of his fame.

WARD, Elizabeth Stuart Phelps (1844-1911), wrote fictional dialogues about heaven: *The Gates Ajar* (1868); *Beyond the Gates* (1883); *The Gates Between* (1887); and *Within the Gates* (1901); which were extremely popular, probably because she pictured heaven as a place where humans might like to be, in contrast to the conventional abstract pictures given by the clergy. She also wrote religious poetry, and several novels that reflect a liberal concern for social justice: in *The Silent Partner* (1871), she pleads the cause of the New England mill girls; *A Singular Life* (1894) pictures a young minister whose parallels to Christ make him unsuitable in the eyes of the orthodox clergy.

WARD, Nathaniel (c. 1578-1652), a British lawyer, minister, and wit, was educated at Cambridge, practised law, and after becoming a minister served British merchants in Prussia. In 1633 he was dismissed from an English post for preaching Puritanism, and he came to Massachusetts, where he lived until 1648, when he returned to England. He wrote the colony's first code of law, "The Body of Liberties," and composed some verse, but he is remembered principally for *The Simple Cobler of Aggawam* (1647), a crotchety argument for political and religious freedom, in theory, which opposed freedom in practice.

WARNER, Charles Dudley (1829-1900), graduated from Hamilton College. After surveying in Mississippi and practising law in Chicago, he became editor of the Hartford, Connecticut *Courant* (1861), for which paper he wrote the Irving-like essays which composed his first book, *My Summer in a Garden* (1871). Several other collections of his reminiscent or critical essays were published, including *Backlog Studies* (1873), *Being a Boy* (1878), and *Fashions in Literature* (1902). With his neighbor, Mark Twain, he wrote *The Gilded Age* (1873), and he later wrote a trilogy portraying the accumulation, misuse, and dissipation of a great fortune, which in some ways anticipated Dreiser: *A Little Journey in the World* (1889), *The Golden House* (1894), and *That Fortune* (1899). Warner also wrote biographies of *Captain John Smith* (1881) and *Washington Irving* (1881) for the "American Men of Letters Series."

WARREN, Mercy Otis (1728 - 1814), reared by a family of ardent liberals, and married to a future Revolutionary general, conducted a salon of Revolutionary politics that was visited by Jefferson, Winthrop, and Adams. She also aided the rebel cause with two plays: *The Adulateur: A Tragedy* (1773); and *The Group* (1775), which satirized Governor Hutchinson and the Tories. Her *History of the Rise, Progress, and Termination of the American Revolution* (3 vols., 1805) is still regarded as an important source not only of the facts but of vital personality sketches. She also wrote *Poems Dramatic and Miscellaneous* (1790). *The Motley Assembly* (1779), a farce concerning Boston Tories, is attributed to her.

WARREN, Robert Penn, (1905-), grew up in Kentucky, took his B.A. at Vanderbilt University where he was a member of *The Fugitive* staff, then held fellowships at California, where he took his M.A., and at Yale. He was a Rhodes Scholar at Oxford where he took the bachelor of letters degree in 1930. For many years he was a profesor of English at Louisiana State University and managing editor of *The Southern Review*. He is presently a professor at Yale. His poems, constructed in the contemporary "neo-metaphysical" tradition of tight and intellectual prosody, have been published in *Pondy Woods and Other Poems* (1930), *Thirty-six Poems* (1935), and *Selected Poems* (1944). His novel, *Night Rider* (1939), was written under the Houghton Mifflin fellowship. *At Heaven's Gate* (1943) portrays a Southern financier. He was awarded the Pulitzer Prize for *All the King's Men* (1946), the story of a political tyrant, presumably reflecting the career of Huey Long.

WATERS, Frank (1902-), grew up in Colorado Springs, in the mining camp of Cripple Creek, Colorado, and on a Navajo reservation. At Colorado College he majored in physics, and while working later for the Pacific Telephone and Telegraph Company as an engineer, he thoroughly covered the West, always taking a scholarly interest in western

legends and history, and in the contrast between the Indian's ceremonial magic and modern scientific magic. His novels include *Pitch* (1930), *The Wild Earth's Nobility* (1935), *Below Grass Roots* (1937), and *The Dust Within the Rock* (1940), the last three forming a trilogy of the mining industry in Colorado. A later trilogy treats minority problems in the Southwest: *People of the Valley* (1941), *The Man Who Killed the Deer* (1942), and *The Yogi of Cockroach Court* (1947). His non-fiction books include *Midas of the Rockies* (1937), *The Colorado* (1946; "Rivers of America" series), and *Masked Gods* (1950).

WEAVER, John V[an] A[llstyn] (1893-1938), writer of light verse in American slang, was born in Charlotte, North Carolina. After Hamilton and Harvard, he worked in varied capacities in various parts of the country, then published *In American* (1921) and similar volumes of verse cleverly using the jargon of the average shirtsleeve worker. His novels and plays, cased in the same language of the street and factory, are humorous and virile yet tenderly sentimental. An autobiography in verse is *Trial Balance* (1931).

WEBSTER, Jean (1876-1916), New York author of juvenile novels, the most famous being *Daddy-Long-Legs* (1912), a sentimental and happy story told through the letters of a young girl to her guardian, whom she eventually marries.

WEBSTER, Noah (1758-1843), Connecticut lawyer, lexicographer, and philologist, graduated from Yale in 1778, and while teaching school to support himself during his study for the law, recognized the dearth of American textbooks. His famous *Spelling Book* was the first part of *A Grammatical Institute of the English Language* (1783-5), which included a grammar and a reader. The speller sold 75,000,000 copies and helped to standardize American pronunciation and spelling. His *Compendious Dictionary of the English Language* (1806) was an apprentice task for his more scholarly *An American Dictionary of the English Language* (2 vols., 1828), which used American as well as English usage as a criterion for selection and became the recognized authority. He was also active as a Federalist in journalism and politics.

WEIDMAN, Jerome (1913-), a New York novelist who satirized the business mores of his city in *I Can Get It For You Wholesale* (1937), and *What's in It for Me?* (1938). His stories were collected in *The Horse That Could Whistle Dixie* (1939). *Too Early to Tell* (1946) satirizes government bureaucracy during the war.

WELTY, Eudora (1909-), Mississippi author, has twice won the O. Henry Memorial award for her stories, and has published collections of short stories that include *A Curtain of Green* (1941), concerned with abnormal, grotesque characters; *The Wide Net* (1943), portraying the disintegration of southern aristocracy; and *The Golden Apples* (1949). Her novelette, *The Robber Bridegroom*

(1942), is a fairy-telling of an old ballad about a gentleman by day who is an outlaw by night, and of his wooing Rosamund, a planter's daughter. *Delta Wedding* (1946) recalls three generations of a plantation family, and *Music from Spain* (1949) pictures a Misissippian in San Francisco.

WESCOTT, Glenway (1901-), grew up in Wisconsin, studied for two years at the University of Chicago, and lived mainly abroad, in Mexico, Germany, and France, until 1934 when he settled in New Jersey. His poems have been published as *The Bitterns* (1920), and *Natives of Rock* (1925). His novels include *The Apple of the Eye* (1924), a portrait of a western boy torn between family duty and natural beauty; *The Grandmothers* (1927) concerning the frustrations of Midwestern life; *The Pilgrim Hawk* (1940), treating love in Paris; and *Apartment in Athens* (1945), chronicling underground resistance to the Nazis. He has also published collections of his short stories and essays.

WESTCOTT, Edwards Noyes (1846-98), a successful Syracuse, New York banker who wrote the posthumously published successful novel, *David Harum, A Story of American Life* (1898), concerning a shrewd and humorous country banker.

WHARTON, Edith [Newbold Jones] (1862-1937), like Henry James, was born to a family of private income which allowed extensive travel, and she was educated abroad as much as in her native New York, usually by a governess. She attempted a novel at 11 and another

at 15, and Longfellow introduced her
juvenile poetry to the pages of the *Atlantic Monthly*. Her life of leisure was not
substantially changed by her marriage
to Edward Wharton, a Boston banker out
of Virginia: they lived in Newport,
Lenox, and abroad, settling permanently
in France as his health declined. After
The House of Mirth (1905) became a
best-seller, she won many literary friends,
including Howells, Henry James, Hardy,
and Meredith. She would bundle James
into her carriage for rides, insisting that
he did not get enough fresh air. Apparently both officious and gracious, she
continually entertained a houseful of
guests, probably doing her writing in the
morning and leaving the manuscripts to
the care of her secretaries, for there was
no "show" of her being a writer. She
modeled her best work on that of Henry
James, striving to perfect her architecture and to give depth to her studies of
high society. Since high society has a
paucity of depth, dramatic significance
can be achieved only through its victims, who are victimized, in the Wharton
novel, simply because they do have depth
as well as social standing. Such a victim
is Lily Bart who gains entrance to *The
House of Mirth* because of her personal
charm and family background and in
spite of her relative poverty. Reluctant
to gain financial security on the social
marriage mart, she is trapped into the
peril of an even less respectable commerce, and escapes through suicide.
Ethan Frome (1911) is based on the
same problem of a sensitive individual
unsuccessfully fighting convention, but
in contrast to most of Mrs. Wharton's

work the setting is a New England farm
with Puritan rather than high society
conventions. Ethan and his love fail to
escape his wife by suicide when their
planned accident misfires and leaves
them her invalid captives. Mrs. Wharton
wrote the novel originally in French.
Despite the far departure from her own
environment, this novel remains her best
work, both for its structure and its powerful development of a tragedy that is announced in the opening pages. *The Age
of Innocence* (1920), which received
the Pulitzer Prize, is her closest parallel
to James, even having for the protagonist, Newland Archer, a Jamesian name.
It concerns Archer's love for a Polish
countess, which is frustrated by conventions. These are the best of her two
dozen novels and novelettes, which also
include *Sanctuary* (1903), a study in
heredity; *The Valley of Decision* (1902),
concerning a liberal Italian aristocrat
of liberal sympathies; *The Custom of the
Country* (1913); *False Dawn* and *The
Old Maid* (1924), both novelettes; and
Hudson River Bracketed (1929), a contrast of Middle Western with New York
culture. Her eleven collections of short
stories include *The Greater Inclination*
(1899), *Tales of Men and Ghosts*
(1910), *Xingu and Other Stories* (1916)
and *The World Over* (1936). Her poems
were collected as *Artemis to Actaeon
and Other Verse* (1909), and *Twelve
Poems* (1926). She was awarded an
honorary doctorate degree by Yale, and
the French government awarded her the
Cross of the Legion of Honor for her
energetic assistance to refugees of the
First World War.

WHEELOCK, John Hall (1886-), New York poet who wrote *Verses by Two Undergraduates* (1905) with Van Wyck Brooks while an undergraduate at Harvard. His post-graduate poems, exuberant and romantic, were published in *The Human Fantasy* (1911) and *The Belovéd Adventure* (1912). Later volumes, such as *The Black Panther* (1922) and *The Bright Doom* (1927) are touched with a mystical philosophy.

WHIPPLE, Edwin Percy (1819-86), a Boston critic and successful lyceum lecturer was one of the most prominent book reviewers of his day. As a critic his estimates of European writers have stood the test of time better than his estimates of American writers, whom he tended to overpraise.

WHITE, E[lwyn] B[rooks] (1889-), has, with his wife, Katherine S. White, been on the staff of *The New Yorker* for many years, and has also written for *Harper's Magazine*. In a period when the informal essay has declined as a literary medium, he has had few rivals in that form, although he leans more heavily on satire than the conventional essayist of earlier periods. His books include *Alice through the Cellophane* (1933); *The Fox of Peapack, and Other Poems* (1938); *Quo Vadimus? or, The Case for the Bicycle* (1939), containing stories and sketches; *One Man's Meat* (1942, enlarged 1944); *Stuart Little* (1945), a story of a mouse; and a collection of his New Yorker editorials, *The Wild Flag* (1946). *Charlotte's Web* (1952) is about a spider.

WHITE, Stewart Edward (1873-1947), grew up in Michigan, where he became familiar with the lumber and logging industry, and in his teens moved to California, where he learned ranch life. After graduating from the University of Michigan he enrolled at Columbia to study law, but was encouraged by Brander Matthews to turn to writing. He has written voluminously—novels, stories, travelogues, and historical studies mainly about outdoor life, either in Michigan or the Far West. His novels include *Arizona Nights* (1904); a trilogy, *The Story of California* (1927); and *Wild Geese Calling* (1940).

WHITE, William Allen (1868-1944), a leader of the Republican party in Kansas and a world famous editor and publisher of the Emporia *Gazette*. He published his independent, outspoken *Gazette* editorials as *The Editor and His People* (1924), and *Forty Years on Main Street* (1937), and wrote several other books on politics. His *Autobiography* (1946) was awarded the Pulitzer Prize.

WHITE, William L[indsay] (1900-), a son of William Allen White, wrote novels and documentary reports of his experiences as a foreign correspondent: *What People Said* (1938), *Journey for Margaret* (1941), *They Were Expendable* (1942), and *Queens Die Proudly* (1943).

WHITMAN, Albery Allson (1851-1901), a Negro poet who wrote in the romantic tradition about Negroes and Indians.

His books include *The Rape of Florida*
(1884) and *An Idyll of the South*
(1901).

WHITMAN, Sarah Helen [Power] (1803
78), is remembered for her defense of
Poe, to whom she was engaged just before his death: *Edgar Poe and His Critics*
(1860). His second "To Helen" was addressed to her, as were his *Last Letters*
(1909). He influenced her poetry also,
collected in *Hours of Life* (1853).

WHITMAN, Walt[er] (1819-92), was
born in Huntington, Long Island, New
York in a farm house that is now only
a few yards from a major traffic artery.
In 1823 his family moved to Brooklyn,
where his father worked as a carpenter,
building and selling houses. A year
later the great Lafayette, in Brooklyn for
a dedication ceremony and helping to
move some small boys out of the way,
picked up young Walt and kissed him.
The family of nine children was extremely poor, and Walt left school in
1830 to work, first as office boy for a
lawyer, and next as a printer's devil for
a newspaper. In 1839 he taught in a
Long Island country school, being engaged in journalism in his spare time,
and by 1841 he had become a full-time
journalist, editing papers both on Long
Island and in New York City. Meanwhile he educated himself in classic
literature, reading Shakespeare, Homer,
Dante, the Nibelungenlied, and even the
Hindu poets. During this period he also
published a few conventional poems
which were not at all precursors of his
future achievements. He also wrote,

merely as a commercial venture, a tem-
perance-propaganda novel, *Franklin
Evans; or, The Inebriate: A Tale of the
Times* (1842), reporting the downward
path of a country boy from the time he
takes his first drink while enroute to
the big and wicked city. Mixed with
Dickensian didacticism are some sur-
prisingly fine scenes. During this period
Whitman was also active in politics and
spoke at Democratic meetings. In 1846
he became editor of the Brooklyn *Eagle*,
a Democratic party paper, and his general behavior in this post reflects his
growing maturity. William Henry Sut-
ton, his errand boy, recalled him many
decades later as "a nice, kind man" who
wore a short beard, dressed conventionally, and carried himself with dignity.
He introduced a column of literature to
the front page (which had previously
contained only advertisements), printing
poems and stories of Bryant, Longfellow, Whittier, Poe, and Hawthorne, as
well as translations of Goethe. His own
editorials were sufficiently provocative
to suggest that the publisher-printer gave
him a free hand: he sympathized with
immigrants and underpaid working girls,
and suggested earlier closing of the
stores to shorten the working day. Following his party line, he ridiculed the
Abolitionists and supported slavery as
an established fact; however, he expressed the hope that it would be abol-
ished gradually with the growth of the
democratic spirit. By 1848 the Demo-
cratic-republican party was split over
the Wilmot Proviso, specifically, over the
question of whether slavery should be
allowed in the new southwestern states.

(241)

The *Eagle* was also split, the publisher taking the southern Democrats' position that slavery should be extended, and Whitman taking the position that it should be checked. Refusing to yield, he was, quite naturally, discharged. (His editorials have been collected as *The Gathering of the Forces,* edited by Cleveland Rogers and John Black, 1920.) In February, 1848, he joined his brother, Jeff, in a trip to New Orleans, where he was on the staff of the *Crescent.* Some scholars have endeavored to prove that he had a love affair with an octoroon, and was fired for this reason, but there is no real evidence to support the story. After a few months he returned to Brooklyn by way of the Mississippi, Chicago, and New York state, and became active in "Free-Soil" politics. He was also a Bohemian dandy for a period, haunting Pfaff's beer cellar and being concerned over the cut of his clothes. For many years he had been an opera enthusiast, and to this former pastime he added the pleasure of riding the horse-drawn omnibuses, becoming a great friend of the drivers. He worked on or edited several papers, including the Brooklyn *Times.* When editorial positions became more difficult to find, he helped his father build cheap houses. But his favorite pastime was simply loafing, soaking life in. He was not strictly a lazy loafer; he embraced leisure philosophically. It was, perhaps, this leisure that was transforming him into a new species of poet; leisure to become a part of a crowd at the beach or in the city, to chat with ferryboat pilots and omnibus drivers, to read Goethe, Hegel,

Carlyle, Emerson, and George Sand's *The Countess of Rudolstadt* which portrayed a great humanitarian poet of the common people. But while he remained a Jeffersonian Democrat of the people, he rebelled more and more against the conventions the people embraced, and in this period his dandy dress was replaced by working garb or any kind of slovenly clothes. The transformations in his mind and personal attitudes during these years will never be interpreted exactly, composed as they are of countless intangibles, but his notebooks reveal that by 1847 he had begun to work on the 12 poems that he eventually shaped to their first of many "perfections" for publication in the first edition of *Leaves of Grass* (1855). He printed the book himself, and had it bound in dark green cloth with gold trim and a rococo cover. Ten pages were devoted to the preface and the 12 untitled poems filled 85 pages. He sent copies to many literary celebrities, and most of them were returned, but Emerson wrote Whitman a letter of high praise—as well he should have, for he had asked for many of the poetic qualities Whitman was now delivering. The reviews, however, were hostile, criticizing the book for its lack of gentility, experimental prosody, and "Transcendental bombast." Scholarly estimates of the number of copies sold now range from 5 to 1500. The second edition (1856) included 20 new poems, with a quotation from Emerson's letter on the back. Much has been made of Emerson's shock at this public display of a private communication, and it is reported that some of the new poems shocked him, but it is probable that he merely had a few misgivings. The third edition (1860) was issued by a reputable Boston publisher, and Whitman spent three months in Boston supervising the publication. Emerson, whom he met frequently, tried to persuade him to omit his too open treatment of sex in the 124 new poems he was including, but Whitman remained self-reliant. There were extensive revisions of the earlier poems and many new titles, and for the first time he arranged the work in the allegorical-dramatic order that became his continuing purpose. In the following years he published eight further editions, each longer than the last and revealing changes in earlier work. Usually the only favorable reviews were those he wrote himself, anonymously. In 1863 he moved to Washington where he acted as a "one-man U. S. O." and voluntary nurse in the over-crowded army hospitals, supporting himself by part-time work. He remained there after the war, working first in the Department of the Interior and later in the Attorney General's office after Interior Secretary Harlan was shocked to learn that this was the author of that immoral *Leaves of Grass,* and fired him. In 1873, suffering from partial paralysis, he took up his residence in Camden, N. J., where he lived out his life except for visits to Colorado (1879) and to Canada (1880). Meanwhile Rossetti had brought out an edition of his poems in England (1868) and he won many English admirers, including Swinburne, Symonds, Stevenson, and a woman named Anne Gilchrist,

who fell in love with him through the reading of his poems and came to Camden seeking his hand in marriage, but he regarded her only as his great friend. In these later years he was truly "the good gray poet," in fact as well as in pose. His poetry leaned more on indirection than on realism, and the individualism which had earlier changed to a fervent nationalism now grew to a spiritual internationalism. *Leaves of Grass* no longer seems immoral, and many critics now rate Whitman as the greatest of American poets. Certainly he was the greatest poetic phenomenon ever to strike the American scene, hailing the American way with greater enthusiasm than a panel of senators on Independence day, proclaiming the freedom and the dignity of the common man or woman in a free verse that used the common idiom. And the reply of the great American people was indifference or hostility. Fortunately, he was too egotistical to become bitter. He knew he was a true prophet, and that prophets are not without honor. The poetry itself, however new and rebellious in form, was romantic, as his imitators in the 20th century are romantic. It differed from the New England school of gentility in presenting an American instead of an English romanticism. An enumeration of some of his theories would sound slightly neo-classical: the poet must perceive the permanent in the transient; style should be natural, avoiding the ornaments of rhyme and meter; the parts of a poem should form a harmonious proportion to the whole. But his attitudes, from the sentimental to the ideal, and his explosive expressions of them, were the romanticism of an immature personality singing the youth of a country. Whereas his earliest and most romantic poems were favored formerly by English professors, the later tendency has been to praise his greater architectural performances, "Crossing Brooklyn Ferry," "Out of the Cradle Endlessly Rocking," and "When Lilacs Last in the Dooryard Bloom'd." He was, finally, romantic in his love of people and the experience of being a person, alive. His *Complete Writings* (10 vols., 1902) were published by his executors, and of the many recent editions of his *Leaves of Grass,* the one edited by Emory Holloway (1948) is recommended for its complete coverage and concise notations of the progressive changes through 11 editions.

WHITTIER, John Greenleaf (1807-92), the son of Quakers, grew up on the Massachusetts farm that is described in his poem, *Snow-Bound* (1866). He had little formal education, but he read every book he could find in the region, and was especially fond of Burns, whom he saw as a parallel to himself in his poetic interpretation of a rural scene. William Lloyd Garrison and Abijah Thayer became interested in Whittier as the result of the poems he submitted to their journals, and were instrumental in arranging for him to study at the Haverhill Academy (1827), and later to enter the profession of journalism. In 1831 his first book was published, *Legends of New-England in Prose and Verse,* followed by *Moll Pitcher* (1832), a pamphlet poem about the legendary heroine of the Revolu-

tionary War. In 1833 his association with Garrison finally resulted in his becoming an ardent Abolitionist, and in addition to writing such pamphlets as *Justice and Expediency* (1883), he entered politics, was elected to the Massachusetts legislature, and founded the Liberty party. He edited anti-slavery papers in New York and Philadelphia, and in small Massachusetts cities. Meanwhile, he continued writing poems. His anti-slavery verse was collected in *Voices of Freedom* (1846). Poems continuing in the rural tradition were collected in *The Chapel of the Hermits* (1853); *The Panorama and Other Poems* (1856), containing the famous "Maud Muller" and "The Barefoot Boy"; and *Home Ballads* (1860). After the Civil War and the end of his political interests, he devoted himself almost exclusively to poetry, and was revered as one of the most popular poets in America. He remains, in our literary history, as a voice of the mid-19th century, a popular balladist who reflected the taste of his day.

WIGGIN, Kate Douglas (1856-1923), a San Francisco novelist whose *Rebecca of Sunnybrook Farm* (1903) was loved by many children of an earlier period.

WIGGLESWORTH, Michael (1631-1705), was the first of a continuous chain of American ministers who have become best-sellers by sugar-coating their sermons with the devices of literature. *The Day of Doom,* a frightening exposition of Calvinist doctrine in doggerel ballad form, pictures the judgment day and the appearance of "the Son of God most dread." Wigglesworth relaxed slightly from the doctrine that babes who die in infancy without baptism must go to Hell. He, at least, assigned them "the easiest room in Hell."

WILBUR, Richard (1921-), is considered by many critics to be the best of the younger poets. He was educated at Amherst, and at Harvard, where he later became a professor of English. In 1952 he was a Guggenheim Fellow, working in New Mexico. He has won the Oscar Blumenthal prize and the Harriet Monroe awards from *Poetry* magazine. His collections of poems include *The Beautiful Changes* (1947) and *Ceremony* (1950).

WILDER, Thornton [Niven] (1897-), accompanied his family from Wisconsin to Hong Kong (1906) where his father became consul general, and later to Shanghai. He returned to the U. S. in 1914 for study in California and Ohio. After serving with the Coast Artillery he took his B.A. at Yale, studied in Rome, then took his M.A. at Princeton. His first novel, *The Cabala* (1926) portrayed a decadent nobility in post-war Italy. *The Bridge of San Luis Rey* (1927), a story of 5 characters providentially finding death when a bridge collapses in Peru, was an immediate best-seller and received the Pulitzer Prize. *The Woman of Andros* (1930) is a study in optimism based on the *Andrea* of Terence. *Heaven's My Destination* (1935) is a comic story of a salesman whose practical application of Christian

principles leads to trouble. *Ides of March* (1948) is a study of Julius Caesar presented through fictional diaries and letters. Wilder had written many one-act plays before he won his second Pulitzer Prize with *Our Town* (1938), a sympathetic study of people in a small New England town. *The Skin of Our Teeth* (1942), which also took the Pulitzer Prize, is an allegory of mankind from the Ice Age to destruction.

WILEY, Calvin Henderson (1819-1887), native North Carolina novelist and educator, wrote two historical romances, *Alamance* (1847) and *Roanoke* (1849), both dealing with Revolutionary times in North Carolina. *Roanoke* was issued as *Life in the South* (1852), and pirated in London as *Utopia* (1851) and also as *Adventures of Old Dan Tucker* (1852). Wiley's stories are unique in his region for the imposition of local traditions and legends on the background of historical events and settings.

WILLIAMS, Ben Ames (1889-1953), published more than 500 short stories and 30 novels, ranging from deliberately constructed best-sellers such as *Leave Her to Heaven* (1944), to the serious historical study of a *House Divided* (1947), for which he did research over a period of 30 years. He grew up in Mississippi, Ohio, and Cardiff, Wales, where his father was United States Consul. After graduating from Dartmouth (1910) he taught himself the craft of writing, working on short stories while reporting for a Boston newspaper by day. Eventually he made millions with his pen, but remained a generous, friendly, and "big" man.

WILLIAMS, John (1761-1818) emigrated from England to America in 1797 out of the necessity created by his scurrilous satires on both political and literary matters (Macaulay called him a malignant and filthy baboon). He continued in this pattern while writing for American journals, using the pseudonym of Anthony Pasquin. His *Hamiltoniad* (c. 1804) is a verse satire which attacks the Federalists.

WILLIAMS, Roger (c. 1603-83), the son of a London tailor, became a protégé of the great jurist, Sir Edward Coke, and received his education at Cambridge. During the evolution of his theological beliefs from the Church of England to a total break with all creed, he came to Boston (1630) in his Puritan period, but was banished for heresy (1635), and after living with the Indians for a year, founded the independent colony of Rhode Island. His beliefs in a democratic theology are clearly and forcefully stated in *Mr. Cotton's Letter Lately Printed, Examined, and Answered* (1644); *The Bloudy Tenent of Persecution* (1644); and *The Bloudy Tenent Yet More Bloody* (1652); all in controversy with John Cotton, who had written the charges that caused Williams to be banished from Massachusetts. Williams was the first militant democrat, in belief and action, on soil where democracy would some day flourish.

WILLIAMS, Tennessee (1914-), the pen name of Thomas Lanier Williams, who was born in Mississippi and spent some of his adolescent years in St. Louis, the setting for his first successful play, *Thé Glass Menagerie* (1945). This sensitively portrayed story depicts a young man breaking from the dream world of his mother, who attempts to maintain her southern gentility in a slum apartment, and from his crippled sister who finds love only in her menagerie of glass animals. The production, with Laurette Taylor as the mother, immediately made Williams famous, and brought a new quality of poetry to the contemporary stage. *A Streetcar Named Desire* (1947) again pictured a heroine's dream world in contrast to the harsh, animal life of her sister and brother-in-law in a New Orleans slum. These were followed by *Summer and Smoke* (1948), *The Rose Tattoo* (1950), and *Camino Real* (1953). *27 Wagons Full of Cotton* (1946) is a collection of his one-act plays.

WILLIAMS, William Carlos (1883-), New Jersey poet and physician, attended schools in Rutherford, N.J., in Switzerland, and in New York; took his M.D. degree at the University of Pennsylvania (1906); and studied pediatrics at Leipzig. Meanwhile he had begun writing poems, and revealed himself an Imagist in *Poems* (1909) and *The Tempers* (1913). In his following volumes he gave more attention to structure, and was perhaps conditioned by his professional services among the proletariat to use the images of common and homely experience in vivid, objective apprehensions of sensory experience. This approach becomes apparent in *Al Que Quiere!* (1917), *Kora in Hell* (1920), *Sour Grapes* (1921), and *Spring and All* (1922). His other volumes include *Collected Poems* (1934), *Complete Collected Poems* (1938), *The Broken Span* (1941), *The Wedge* (1945), and *Paterson,* (4 books, 1946-) a long poem study of the New Jersey city where he lives, of which two volumes have been issued to date. He has also written novels, including *A Voyage to Pagany* (1928), *White Mule* (1937), and *In the Money* (1940). His short stories have been collected as *The Knife of the Times* (1932) and *Life along the Passaic River* (1938), and his impressionistic essays as *The Great American Novel* (1923) and *In the American Grain* (1925).

WILLIAMSON, Thames [Ross] (1894-), ran away from his Idaho home at 14 to become a hobo, sheepherder, cabin boy, prison official, and reporter. After graduating from the University of Iowa (1917) he took his M.A. at Harvard, on a fellowship, in economics and anthropology, and studied for but did not complete the Ph.D. He taught economics at Simmons and Smith Colleges, and wrote textbooks on the social sciences, their successful sales finally enabling him to devote his full time to writing novels which reflect his hard life as a boy: *Run, Sheep, Run* (1925), *Gypsy Down the Lane* (1926), *The*

Man Who Cannot Die (1926), *Stride of Man* (1928), *Hunky* (1929), *The Woods Colt* (1933), and *Under the Linden Tree* (1935). Disappointed over the reception of his novels (although two were Book-of-the-Month Club selections), he became a Hollywood scenarist.

WILLIS, N[athaniel] P[arker] (1806-67), New York journalist, dramatist, versifier, traveling correspondent, and short story writer, tried whatever avenue would lead to the highest rate of pay. As an editor he contracted to publish a column by Thackeray, then little known, and also employed Poe on the *Mirror.* In his own day he enjoyed an unwarranted reputation and was the darling of literary society.

WILSON, Edmund (1895-), was a classmate of Fitzgerald at Princeton, with whom he wrote a libretto for a Triangle Club production. He served overseas (1917-19) during the War, and upon his return became engaged in criticism and editing; serving as managing editor of *Vanity Fair,* as a staff member of *The New Republic,* and as literary critic of *The New Yorker.* An "artist's advocate," a critic of criticism, and a rebel against the platitudes of art and society, Wilson has been something of a lone wolf in American letters. His critical studies have been published as: *Axel's Castle* (1931), an early study of Symbolism and other imaginative literature from 1870 to 1930; *The Triple Thinkers* (1938); *The Wound and the*

Bow (1941); *The Boys in the Back Room* (1941), concerning California writers; and *Classics and Commercials* (1950). *I Thought of Daisy* (1929) is a novel, and *Memoirs of Hecate County* (1946) contains six sophisticated stories of New York life including a damning satire on the confusion of taste with commerce in book club activities.

WINTERS, [Arthur] Yvor (1900-), poet, critic, and teacher, grew up in California and Washington, and attended high school in Chicago. After two years at the University of Chicago (1917-19) he was compelled to spend three years in a Santa Fe sanitarium because of tuberculosis. He took his B.A. and M.A degrees at the University of Colorado, and his Ph.D. at Stanford, where he is now a professor of English. His poems have been published as *The Immobile Wind* (1921), *The Bare Hills* (1927), *The Proof* (1930), *The Journey* (1931), *Before Disaster* (1934), *Poems* (1940), and *Giant Weapon* (1944). His rather classical poems have been less provocative, however, than his criticism, which has been over-praised and over-damned, and seldom objectively studied. As with Pound, although not to so violent a degree, Winters' personality constantly emerges to distract the reader from his involved analyses. Like most contemporary critics, he prefers the English neo-classical poetry to 19th-century romanticism, but he sees the deterioration as beginning with the 18th century. His moral and critical absolutism have led him to a theistic position, as with many of his contemporaries, yet he agrees with few of them in the poetry he selects to admire. The major publications of his criticism are *Primitivism and Decadence* (1937), *Maule's Curse* (1938), and *Anatomy of Nonsense* (1943), which are collected with other work as *In Defense of Reason* (1947). His other major work is a study of *Edwin Arlington Robinson* (1947).

WISTER, Owen (1860-1938), Pennsylvania novelist, spent much of his boyhood abroad, and after graduation from Harvard, went to Paris to study musical composition. Later abandoning his musical career, he graduated from the Harvard Law School (1888), and became a Philadelphia lawyer. Meanwhile he had made several trips to Wyoming for his health, and finding that his sketches of western life were successful, turned to writing. His stories were collected as *Lin McLean* (1898), and *The Jimmyjohn Boss* (1900). His novel, *The Virginian* (1902), which added touches of domesticity to the conventional western story, became a best-seller. *Lady Baltimore* (1906), a romantic novel set in Charleston, prompted a 5000-word letter from his Harvard classmate, Theodore Roosevelt, the subject of a biography by Wister published in 1930. He also wrote biographies of Grant (1900) and Washington (1907), and a charming whimsical story of two college students studying for *Philosophy 4* (1903).

WOLFE, Thomas [Clayton] (1900-1938), born in Asheville, North Carolina, began

with *Look Homeward, Angel* (1929) a
series of partially autobiographical novels
noted for their verbal richness as well
as their seeming lack of form. Actually
they minimize plot in the usual sense of
the word, following instead the career of
a sensitive artistic youth in conflict with
an unsympathetic world. *Look Homeward, Angel* relates the childhood of
Eugene Gant (Wolfe himself) in the
mountain town Altamont (Asheville) in
Old Catawba (North Carolina) and his
undergraduate years at the state university at Pulpit Hill (the University of
North Carolina at Chapel Hill). The
"buried life" of the hero is contrasted
with his outer life. *Of Time and the
River* (1935) takes Gant to Harvard
where he writes plays, then to New York
where he finds no producers for them
and turns to teaching English at a
city institution (New York University),
and finally to adventures in Europe.
In the first posthumous novel, *The Web
and the Rock* (1939), Wolfe renames
his hero George Webber, gives him a
different physique and background, but
soon returns to semi-autobiography by
relating a tortured love affair with a
talented older Jewish woman. *You Can't
Go Home Again* (1940) concludes the
tetralogy; Webber lives in Brooklyn
during the Depression, becomes a successful novelist, breaks with his publisher
Foxhall Edwards (Maxwell Perkins of
Scribner's), and the book ends on a note
of triumphant belief in mankind's future.
Wolfe has been admired particularly for
his clear presentation of an individual's
spiritual isolation in a hostile environment, for the opulent and profuse poetry
of his prose, and for his emotional,
mystical, and intense Americanism.
Some readers have been repelled by
what they consider his egocentricity.
Yet Wolfe remains one of the strong-
est and most original contemporary
writers and seems destined to speak
of the enchantments and torments of
youth to each generation. His short
pieces are collected in *From Death to
Morning* (1935) and *The Hills Beyond*
(1941). *The Story of a Novel* (1936) is
a frank revelation of his creative experiences. There are two volumes of his
poetic passages, and varied collections
of his letters and journals. *Mannerhouse*
(1948) is an early play, suitable for
studying his later style and characters.
Wolfe died of a cranial operation at
Johns Hopkins Hospital, just short of his
thirty-eighth birthday.

WOODBERRY, George Edward (1855-
1930), a conservative critic, poet, scholar,
and professor (at Columbia), wrote biographies of Poe (1885), Hawthorne
(1902), and Emerson (1907); essays
collected in *Heart of Man* (1899), *Mak-
ers of Literature* (1900), *The Torch*
(1905), and *The Appreciation of Literature* (1907); as well as several books
of poems.

WOOLLCOTT, Alexander [Humphreys]
(1887-1943), after graduation from
Hamilton College (1909), became a
New York drama critic. In 1929 he
became a radio reviewer of books, and
the "Town Crier." His gossipy essays
published in *While Rome Burns* (1934)

sold more than a quarter million copies. With George S. Kaufman he wrote two plays, *The Channel Road* (1931) and *The Dark Tower* (1933). He also acted in several plays, including *The Man Who Came to Dinner* (by Kaufman and Hart, 1939), in which he played the part of himself.

WOOLMAN, John (1720-72), the Quaker humanitarian, began life as a tailor's apprentice, baker, and shopkeeper. At 26 he became an itinerant preacher, and although untrained for the ministry, he was better suited for the preaching of Christianity than many who graduated from the theological schools. He cried for justice to both Indian and slave, deprecated riches and war, and advocated an equitable distribution of the world's goods. His works were written in a clear style, and include *Some Considerations on the Keeping of Negroes* (1754), *A Plea for the Poor* (1763), and his *Journal* (1774) of his inner life. He was highly praised by Lamb in England and by Whittier in America.

WRIGHT, Richard (1909-   ), grew up in Natchez, Mississippi with little formal education, and first attracted attention when he received the *Story* Prize for his collection of four stories of southern prejudice against his race, *Uncle Tom's Children* (1938). *Native Son* (1940) is about a Negro boy growing up in the Chicago slums and murdering two women out of confusion. *Black Boy* (1945) is an autobiographical novel.

WRIGHT, Willard Huntington (1888-
1939), editor of *Smart Set* (1913-14), student of *Modern Painting* (1915) and critic of *The Future of Painting* (1923), wrote *Europe after 8:15* (1913) in collaboration with Mencken and Nathan, and *What Nietzsche Taught* (1914). While recovering from a physical breakdown, and ordered by his physician to restrict his reading to very light novels, such as mystery stories, he invented a sleuth as sophisticated as himself, Philo Vance, who solved *The Benson Murder Case* (1926) and a series of others that became highly popular in their day. In publishing them, Wright used the pseudonym, S. S. Van Dine.

WYLIE, Elinor [Hoyt] (1885-1928), was born to a prominent Philadelphia family. After an unfortunate marriage she eloped with Horace Wylie to England, where she privately and anonymously published her first volume of poems, *Incidental Numbers* (1912). After her return to America she published a collection of polished poems, *Nets to Catch the Wind* (1921), followed by *Black Armour* (1923), both showing a metaphysical influence. Her delicately polished novels include *Jennifer Lorn* (1923), *The Venetian Glass Nephew* (1925), and *The Orphan Angel* (1926), which pictures Shelley being rescued from drowning and brought to America. In 1923 she married William Rose Benét, who edited her *Collected Poems* (1932) and *Collected Prose* (1933).

WYLIE, Philip (1902-   ), left Princeton after three years and became a public relations expert in New York, and later joined the staff of *The New Yorker*. He became a student of the psychology of Jung, whose theories permeate *Genera-* *tion of Vipers* (1943), written in eight weeks; and his *Essay on Morals* (1947). *Opus 21* (1949) indicates the number of Wylie's books up to that year.

Y

YOUNG, Stark (1881-), grew up in Mississippi, and after attending the State University, took his M.A. at Columbia in 1902. He later taught English at the Universities of Mississippi and Texas, and at Amherst. In 1921 he became an editor of *The New Republic* and *Theatre Arts Monthly,* and directed plays for the Theatre Guild. His publications include *The Blind Man at the Window* (1906), a collection of poems; a verse play, *Guenevere* (1906); essays on *The Flower in Drama* (1923); *The Colonnade* (1924), and *The Saint* (1925), plays; short stories collected as *The Street of the Islands* (1930) and *Feliciana* (1935); and the novels, *Heaven Trees* (1926), *The Torches Flare* (1928), *River House* (1929), and *So Red the Rose* (1934).

Z

ZATURENSKA, Marya (1902-   ), was born in Russia but came to America in her youth and attended the University of Wisconsin. She was awarded the Pulitzer Prize in 1938 for her collection of poems, *Cold Morning Sky*. With Horace Gregory, her husband, she wrote a *History of American Poetry 1900-1940* (1946).